INTERNATIONAL CENTRE FOR MECHANICAL SCIENCES

COURSES AND LECTURES - No. 212

WILLIAM PRAGER

BROWN UNIVERSITY, PROVIDENCE, R. I.

INTRODUCTION TO STRUCTURAL OPTIMIZATION

COURSE HELD AT THE DEPARTMENT
OF MECHANICS OF SOLIDS
OCTOBER 1974

UDINE 1974

SPRINGER-VERLAG WIEN GMBH

ISBN 978-3-211-81291-4   ISBN 978-3-7091-2644-8 (eBook)
DOI 10.1007/978-3-7091-2644-8

This text has been submitted as being "ready for camera" by Professor W. Prager and has been reproduced without any corrections or additions, except for the titles at the top of pages.

For the CISM
Dr. Giuseppe Longo
Responsible for the Editorial Board

PREFACE

The six chapters of this <u>Introduction to Structural Optimization</u> correspond to six lectures given at the International Centre for Mechanical
Sciences in Udine in October 1974, as part of a course on structural optimization, the other lecturers being Professors P. Brousse, A. Cyras,
G. Maier, and M. Save.

The first three chapters are concerned with the derivation of necessary
and sufficient conditions for global optimality from extremum principles
of solid mechanics. Whereas the layout of the structure that is to be
designed is regarded as given in these chapters, the remaining three
chapters are concerned with the optimization of the structural layout.

The author feels indebted to Professors W. Olszak and L. Sobrero of the
International Centre for Mechanical Sciences for the invitation that in-
duced him to prepare these lecture notes.

Udine, October 1974   W. Prager

C O N T E N T S

PART I

THE USE OF CLASSICAL EXTREMUM PRINCIPLES IN OPTIMAL STRUCTURAL DESIGN

1. Basic Concepts and Theorems

1.1 Generalized loads and displacements. In structural theory, the members of a structure are not, in general, treated as three-dimensional
continua but rather as continua of one or two dimensions. Rods, beams,
and arches are representatives of the first class, and disks, plates,
and shells, of the second.

A point of a one- or two-dimensional member may respectively be specified by a single parameter or a pair of parameters. The letter x will be
used to denote these parameters. For an arch, for instance, x may stand
for the arc length between the considered point and a reference point on
the arch, and dx will be used to denote the line element of the arch.
For a spherical shell, on the other hand, x may denote the longitude and
latitude of the considered point, and dx will be used to denote the area
of an element of the shell. For the sake of unified terminology, dx will
be called the volume of the considered element of the arc or shell, and
the term specific will be used in the sense of per unit volume.

To introduce some concepts that will be used throughout these notes, consider a horizontal elastic beam that is built in at the end $x = 0$ and
simply supported at the end $x = \ell$, and subjected to a distributed vertical, downward load of the specific intensity $P_1(x)$ and a distributed
counterclockwise couple of the specific intensity $P_2(x)$. To simplify the
terminology, we shall speak of $P_\alpha(x)$, ($\alpha = 1, 2$), as the generalized
loads acting on the beam. For the sake of brevity, we shall not discuss
concentrated loads and couples in this chapter.

With the generalized loads P_α , we shall associate generalized displacements p_α , which are supposed to be small; they will be defined in such a manner that the specific external work $w^{(e)}$ of the loads P_α on the displacements p_α is given by

$$w^{(e)} = P_\alpha \, p_\alpha \, .\tag{1.1}$$

Here, and throughout these notes, a repeated letter subscript in a monomial implies summation over the range of the subscript unless the contrary is explicitly indicated by the words "no summation". For the beam example, $p_1(x)$ and $p_2(x)$ obviously are the vertical, downward deflection of the centerline and the counterclockwise rotation of the cross section at the abscissa x.

The generalized displacements of a structure are subject to certain continuity requirements. For our beam, for instance, the displacements $p_\alpha(x)$ will be required to be continuous and piecewise continuously differentiable. We shall refer to these requirements as the kinematic continuity conditions. Their general form will be discussed in Section 1.2.

The generalized displacements are moreover subject to kinematic constraints that may be external or internal. For our beam, the external constraints are

$$p_1(0) = 0 \, , \quad p_1(\ell) = 0 \, , \quad p_2(0) = 0 \, .\tag{1.2}$$

Bernoulli's hypothesis, according to which material cross sections remain normal to the material centerline, would impose the internal constraint

$$p_1' + p_2 = 0 \, ,\tag{1.3}$$

where the prime denotes differentiation with respect to x. At present, however, we do not impose this constraint.

Generalized displacements that satisfy the kinematic continuity conditions and kinematic constraints will be called kinematically admissible.

To each kinematic constraint, there corresponds a <u>reaction</u>. For example, the reactions associated with the external constraints (1.2) are vertical forces at $x = 0$ and $x = \ell$, and a clamping couple at $x = 0$. Note that the work of any one of these reactions on arbitrary kinematically admissible displacements vanishes. It will be assumed in the following that all kinematic constraints that are imposed on the structure are <u>workless</u> in this sense. Reactions to internal constraints will be discussed at the end of Section 1.2.

<u>1.2. Generalized stresses and strains</u>. The state of stress at a typical point of a structural member is specified by a number of stress resultants or <u>generalized stresses</u> Q_j . For a beam on which the Bernoulli hypothesis is not imposed, bending moment and shear force may be used as generalized stresses. Alternatively, it may be convenient to use the bending moment Q_1 and the product Q_2 of the shear force by the height h of the beam to have generalized stresses of the same dimension.

The generalized loads and the generalized stresses and their derivatives are connected by the equations of equilibrium. For example, if the second choice above is made for the generalized stresses of our beam and the sign conventions in Fig. 1.1 are adopted, the equations of equilibrium are

$$Q_1' - (Q_2/h) + P_2 = 0 , \quad (Q_2/h)' + P_1 = 0 . \tag{1.4}$$

Since concentrated loads and couples have been excluded, the generalized stresses are continuous and their derivatives appearing in the equilibrium conditions are piecewise continuous with jumps occurring only where the generalized loads or the height of the beam have discontinuities. We shall refer to these continuity requirements for the generalized stresses as the <u>static continuity conditions</u>.

The generalized stresses are moreover subject to <u>static constraints</u>. For our beam, the only static constraint is $Q_1(\ell) = 0$.

Generalized stresses that satisfy the static continuity conditions and
the static constraints will be called <u>statically admissible</u>.

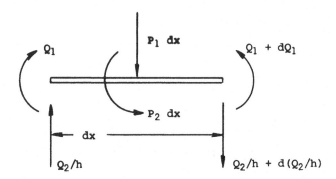

Fig. 1.1: Sign conventions for generalized loads
and stresses of beam element

Once the generalized stresses Q_j have been chosen, the <u>associated generalized strains</u> q_j will be defined in such a manner that the <u>specific internal work</u> $w^{(i)}$ of the stresses Q_j on the strains q_j is given by

$$w^{(i)} = Q_j \, q_j \, . \tag{1.5}$$

To find the expressions of the generalized strains in terms of the generalized displacements and their derivatives, we stipulate that the <u>prin-
ciple of virtual work</u> should hold in the form

$$\int Q_j \, q_j \, dx = \int P_\alpha \, p_\alpha \, dx + W_B \, , \tag{1.6}$$

where the integration is extended over an arbitrary unsupported region of
the structure, and W_B denotes the work of the generalized stresses at the
boundary B of this region on the generalized displacements at B. We then
use the equations of equilibrium to express the generalized loads in the
integral on the right of (1.6) in terms of the generalized stresses and
their derivatives, and remove the latter by integration by parts. Finally,
we obtain the desired expressions for the generalized strains by comparing

the coefficients of the generalized stresses on the two sides of the resulting equation.

Applied to the unsupported region $a \leq x \leq b$ of our beam, these steps are
as follows. Equation (1.6) takes the form

$$\int_a^b (Q_1\, q_1 + Q_2\, q_2)\, dx = \int_a^b (P_1\, p_1 + P_2\, p_2)\, dx$$

$$+ \left(Q_1\, p_2 + (Q_2/h)\, p_1 \right)_a^b , \qquad (1.7)$$

where $\left(f(x) \right)_a^b = f(b) - f(a)$. Substituting P_1 and P_2 from (1.4) and integrating by parts, one obtains

$$\int_a^b (Q_1\, q_1 + Q_2\, q_2)\, dx = \int_a^b \{ Q_1\, p_2' + (Q_2/h)(p_1' + p_2)\}\, dx .$$
$$(1.8)$$

Comparison of the coefficients of the generalized stresses on the left
and right of (1.8) yields

$$q_1 = p_2' , \quad q_2 = (p_1' + p_2)/h . \qquad (1.9)$$

If Bernoulli's hypothesis is adopted,

$$h\, q_2 = p_1' + p_2 \equiv 0 , \qquad (1.10)$$

and Q_2 , which disappears from (1.8), must be regarded as a <u>reaction</u> to
the <u>internal</u> kinematic constraint (1.10) rather than a generalized stress.
Because this constraint implies $q_2 = 0$, the work of this reaction vanishes.

It should be noted that the term W_B in (1.6) vanishes when the principle
of virtual work is applied to the <u>entire</u> beam rather than only to a segment of it. This is a consequence of the kinematic and static constraints
at the ends of the beam.

It is also worth noting that it was not assumed in the preceding discussion that the generalized stresses and strains have the relation of cause and effect. The principle of virtual work only requires that the generalized stresses are statically admissible and that the generalized strains are kinematically admissible, that is, that they are derived from kinematically admissible displacements.

1.3. Elastic stress-strain relations. Strain energy and complementary energy. For a linearly elastic structure, there exists a positive definite specific strain energy

$$e_q = \frac{1}{2} c_{jk} q_j q_k \tag{1.11}$$

such that the specific work of the stresses Q_j on the strain increments dq_j is given by

$$Q_j \, dq_j = de_q = (\partial e_q / \partial q_j) \, dq_j . \tag{1.12}$$

The elastic coefficients c_{jk} satisfy the symmetry relation

$$c_{jk} = c_{kj} ; \tag{1.13}$$

they depend on the elastic constants of the structural material and on the structural dimensions at the considered point, but not on the stresses or strains.

Comparison of the first and last terms in (1.12) furnishes

$$Q_j = \partial e_q / \partial q_j . \tag{1.14}$$

In view of (1.13), it follows from (1.11) and (1.14) that

$$Q_j = c_{jk} q_k . \tag{1.15}$$

Note that (1.11) and (1.15) furnish

$$e_q = \tfrac{1}{2} \, Q_j \, q_j \ .$$ (1.16)

The linear equations (1.15) may be solved for the strains to yield

$$q_j = \bar{C}_{jk} \, Q_k \ .$$ (1.17)

The expression

$$e_Q = \tfrac{1}{2} \, \bar{C}_{jk} \, Q_j \, Q_k$$ (1.18)

will be called the <u>specific complementary energy</u>. Its value equals that of the specific strain energy e_q provided that stresses and strains are related to each other by (1.17).

The integral

$$E_q = \int e_q \ dx = \int C_{jk} \, q_j \, q_k \ dx \ / \ 2 \ ,$$ (1.19)

over the entire structure, of the specific strain energy $e_q(x)$ of a kinematically admissible strain field $q_j(x)$ is called the <u>strain energy</u> of the field $q_j(x)$. Similarly, the integral

$$E_Q = \int e_Q \ dx = \int \bar{C}_{ik} \, Q_j \, Q_k \ dx \ / \ 2$$ (1.20)

of the specific complementary energy $e_Q(x)$ of a statically admissible stress field $Q_j(x)$ is called the <u>complementary energy</u> of the field.

Let $q_j(x)$ and $q_j^*(x)$ be kinematically admissible strain fields that are not identical. It follows from the positive definite character of e_q that the strain energy computed from the difference field $q_j^*(x) - q_j(x)$ is positive:

$$\int C_{jk} \ (\, q_j^* - q_j \,)(\, q_k^* - q_k \,) \ dx > 0 \ .$$ (1.21)

In view of the symmetry relation (1.13) and the definition (1.19), the inequality (1.21) can also be written in the form

$$E_q^* - E_q > \int (q_j^* - q_j) \, Q_j \, dx \, , \tag{1.22}$$

where Q_j is the stress associated with the strain q_j by the stress-strain relation (1.15), and E_q^* and E_q are the strain energies of the fields $q_j^*(x)$ and $q_j(x)$. The inequality (1.22) will be used in Section 1.4. We leave it to the reader to derive the analoguous inequality

$$E_Q^* - E_Q > \int (Q_j^* - Q_j) \, q_j \, dx \, , \tag{1.23}$$

in which Q_j^* and Q_j are statically admissible stress fields that are not identical, E_Q^* and E_Q are their complementary energies, and q_j is the strain associated with the stress Q_j by the stress-strain relation (1.17).

1.4. Extremum principles for linearly elastic structures. Consider a linearly elastic structure with given coefficients of elasticity $C_{jk}(x)$. Let the structure be supported in a way that precludes any rigid-body motion of the entire structure. Given the loads $P_\alpha(x)$ acting on the structure, we wish to determine the displacements $p_\alpha(x)$, the strains $q_j(x)$, and the stresses $Q_j(x)$ caused by these loads. Note that these fields have to satisfy the following conditions:

(1) the displacement field $p_\alpha(x)$ must be kinematically admissible;

(2) the stress field $Q_j(x)$ must be statically admissible; and

(3) the stresses Q_j and the strains q_j, which are derived from the displacements p_α, must satisfy the stress-strain relation (1.15).

It can be shown that there exist fields p_α, q_j, and Q_j satisfying these three conditions, but this existence proof is outside the scope of these notes. Taking existence for granted, we now prove uniqueness. To this end, we first assume that both the fields p_α, q_j, Q_j and $\bar{p}_\alpha$, $\bar{q}_j$, $\bar{Q}_j$ satisfy the three conditions above. Because the structure is linearly elastic, and because both sets of fields correspond to the same loads P_α, the difference fields $\bar{p}_\alpha - p_\alpha$, $\bar{q}_j - q_j$, $\bar{Q}_j - Q_j$ correspond to zero loads. Application of the principle of virtual work to these difference fields and the vanishing loads thus yield

$$\int (\bar{Q}_j - Q_j) (\bar{q}_j - q_j) \; dx = 0 \; . \tag{1.24}$$

By (1.16), the integrand in (1.24) is twice the specific strain energy
for the displacements $\bar{q}_j - q_j$. In view of the positive definiteness
of the specific strain energy, (1.24) requires that

$$\bar{q}_j - q_j \equiv 0 \; . \tag{1.25}$$

The difference $\bar{p}_\alpha - p_\alpha$ must therefore represent a rigid-body displacement. Since this is supposed to be excluded by the supports, the difference $\bar{p}_\alpha - p_\alpha$ vanishes identically. In view of (1.25), the stress-strain
relation (1.15) finally shows that the difference $\bar{Q}_j - Q_j$ vanishes identically. This completes the uniqueness proof.

Let p_α , q_j , and Q_j be the displacement, strain, and stress fields that
represent the solution of our structural problem, and let p^*_α be an arbitrary kinematically admissible displacement field that is not identical
with p_α , and q^*_j the corresponding strain field. Since the stress field
Q_j is statically admissible, application of the principle of virtual
work to the stresses Q_j and the kinematically admissible displacements
$p^*_\alpha - p_\alpha$ and the strains $q^*_j - q_j$ yields

$$\int P_\alpha (p^*_\alpha - p_\alpha) \; dx = \int Q_j (q^*_j - q_j) \; dx \; . \tag{1.26}$$

Use of the inequality (1.22) in (1.26) and rearrangement of terms furnishes

$$E^*_q - \int P_\alpha \, p^*_\alpha \; dx > E_q - \int P_\alpha \, p_\alpha \; dx \; . \tag{1.27}$$

For an elastic structure subjected to given loads P_α , the expression on
the left of (1.27) is called the __potential energy__ for the kinematically
admissible displacements p^*_α and the corresponding strains q^*_j . It is ob-

tained by subtracting the virtual work of the loads on the displacements p_α^* from the strain energy for the strains q_j^* . The inequality (1.27) shows that the displacements and strains of the solution of our structural problem minimize the potential energy (principle of minimum potential energy).

Denoting by Q_j^* an arbitrary statically admissible stress field that is not identical with the stress field Q_j of the solution of our problem, one can show in a similar manner that

$$E_Q^* > E_Q ,\qquad\qquad\qquad (1.28)$$

where the two sides are the complementary energies of the stress fields Q_j^* and Q_j . Thus, the stress field of the solution minimizes the complementary energy (principle of minimum complementary energy).

1.5. Extremum principles for rigid, perfectly plastic structures. The constitutive equations of rigid, perfectly plastic structural elements are usually written in terms of the generalized stresses Q_j and the associated generalized strain rates. Since the strains themselves do not occur in the equations, no confusion will be caused by using the symbol q_j to denote the typical generalized strain rate. Similarly, the symbol p_α will now be used to denote the typical generalized velocity.

It will be convenient to use a Euclidean space in which the stresses Q_j at a point of the structure are represented by the rectangular Cartesian coordinates of the stress point, whose position vector Q will be called the stress vector. Similarly, the vector q with the components q_j will be called the strain rate vector.

The mechanical behavior of a rigid, perfectly plastic structural element is most conveniently specified by its dissipation function D(q) ; this gives the rate at which mechanical energy is dissipated, per unit volume, in plastic flow with the strain rate vector q . The dissipation function D(q) thus represents the specific power of dissipation, which is nonnegative. Since a rigid, perfectly plastic structural element does not exhibit viscosity, the dissipation function is homogeneous of the order one:

$$D(0) = 0 ,$$

$$D(\mu q) = \mu \, D(q) \quad \text{for } \mu > 0 .$$

(1.29)

Furthermore, the dissipation function is supposed to be convex. In view of (1.29), this assumption is expressed by the inequality

$$D(q + r) \leq D(q) + D(r) .$$

(1.30)

The discussion of the thermodynamic foundations of the convexity assumption is beyond the scope of these notes.

According to (1.29) and (1.30), the dissipation function has the properties of the supporting function of a convex domain, called yield domain, whose points have position vectors Q satisfying

$$Q \cdot q \leq D(q) \quad \text{for all } q ,$$

(1.31)

where the center dot indicates scalar multiplication. Interior points of the yield domain represent states of stress below the yield limit for which the element is rigid. Points on the boundary of the yield domain, called yield locus, represent states of stress under which plastic flow may occur. Finally, points outside the yield domain represent states of stress that cannot be attained in the considered structural element.

For a fixed strain rate vector q , the boundary of the half-space (1.31) is a supporting plane of the yield domain. Points Q that this plane and the yield locus have in common represent states of stress under which the

strain rates specified by q may occur. These states of stress will be
said to correspond to the strain rate vector q .

If Q is a state of stress that corresponds to a given strain rate vector q , and Q* is any other state of stress at or below the yield limit,
it follows from (1.31) that

$$(Q - Q^*)\cdot q \geq 0 :$$ (1.32)

the virtual specific power of dissipation computed from an arbitrary
state of stress Q* at or below the yield limit and a given strain rate
q cannot exceed the power computed from this strain rate and a corresponding state of stress Q (principle of maximum local power of dissipation).

The preceding discussion was concerned with the local velocity, strain
rate, and stress. We now consider velocity fields $p_\alpha(x)$, strain rate
fields q(x), and stress fields Q(x) .

A velocity field is called kinematically admissible if it satisfies the
kinematic continuity conditions and constraints for the considered structure. For a rigid, perfectly plastic beam, for instance, on which the
Bernoulli hypothesis is imposed, the rate of deflection must be continuous and piecewise continuously differentiable; moreover, it must vanish
at the supports, and its first derivative must vanish at a built in end.

Consider a rigid, perfectly plastic structure that remains rigid under
the loads P_α . We wish to determine the load factor λ for plastic collapse as follows: plastic flow should be possible under the loads λ P_α but
not under loads λ* P_α when λ* < λ . The fundamental theorems of limit analysis give extremum characterizations of the load factor λ .

The static theorem states that the load factor for plastic collapse spe-
cifies the greatest multiple of the given loads for which there exists a

statically admissible stress field that nowhere exceeds the yield limit. To prove this, denote the greatest multiple of the loads by $\bar{\lambda}$ P and assume that the load factor for plastic collapse has the value $\lambda < \bar{\lambda}$. If the velocities and strain rates of the collapse mechanism under the loads λ P_α are denoted by p_α and q_j , we have

$$\lambda \int P_\alpha p_\alpha \, dx = \int D(q_j) \, dx \ . \tag{1.33}$$

Because there exists a statically admissible stress field $\bar{Q}_j$ for the loads $\bar{\lambda}$ P_α that nowhere exceeds the yield limit, it follows from the principle of virtual work and (1.31) that

$$\bar{\lambda} \int P_\alpha p_\alpha \, dx = \int \bar{Q}_j q_j \, dx \leq \int D(q_j) \, dx \ . \tag{1.34}$$

The contradiction between the assumption that $\lambda < \bar{\lambda}$ and the inequality $\lambda \geq \bar{\lambda}$, which follows from (1.33) and (1.34), establishes the static theorem.

The kinematic theorem states that the load factor for plastic collapse is given by the minimum of the expression

$$\int D(q_j) \, dx \, / \int P_\alpha p_\alpha \, dx \tag{1.35}$$

over all kinematically admissible velocity fields p_α and the corresponding strain rate fields q_j . To prove this, let $\bar{\lambda}$ be the minimum value of the expression (1.35), and let $\bar{p}_\alpha$ and $\bar{q}_j$ be the velocity and strain rate fields that furnish this minimum value. Assume now that the load factor for plastic collapse has the value $\lambda > \bar{\lambda}$. For the loads λ P_α , there then exists a statically admissible stress field Q_j that nowhere exceeds the yield limit. It follows from the principle of virtual work and (1.31) that

$$\lambda \int P_\alpha \bar{p}_\alpha \, dx = \int Q_j \bar{q}_j \, dx \leq \int D(q_j) \, dx \ , \tag{1.36}$$

while from the definition of $\bar{\lambda}$ it follows that

$$\bar{\lambda} \int P_\alpha \, p_\alpha \, dx = \int D(q_j) \, dx \; . \tag{1.37}$$

The contradiction between the assumption that $\lambda > \bar{\lambda}$ and the inequality $\lambda \leq \bar{\lambda}$, which follows from (1.36) and (1.37), establishes the kinematic theorem.

2. Optimal Design of Elastic Beams with Compliance Constraints

2.1. Sandwich beams subject to a single compliance constraint. The method of this section applies to sandwich plates and shells as well as to sandwich beams, but for the sake of brevity only sandwich beams will be discussed in this chapter.

Consider a sandwich beam with segmentwise constant rectangular cross section, and assume that the segment boundaries are given. (The case where these boundaries are at the choice of the designer will be treated in Chapter 4.) The width b_i and the height $2 h_i$ of the core section of the i-th segment are given, but the thickness t_i of the identical cover plates is at the choice of the designer subject to the condition that

$$\bar{t} \le t_i \le \bar{\bar{t}} , \tag{2.1}$$

where $\bar{t}$ and $\bar{\bar{t}}$ are given values. The lower bound $\bar{t}$ is introduced to exclude overly thin cover plates, and the upper bound $\bar{\bar{t}}$, to avoid a violation of the basic assumption that the ratio t_i / h_i is small in comparison to unity.

In analyzing the deflection $p(x)$ of the beam under the given distributed load $P(x)$, we shall adopt Bernoulli's hypothesis. Accordingly, the bending moment $Q(x)$ is the only generalized stress, and the curvature $q(x) = - p''(x)$ is the corresponding generalized strain. In the i-th segment, bending moment and curvature are related by

$$Q(x) = c_i \, q(x) , \tag{2.2}$$

where

$$c_i = 2 \, E \, b_i \, h_i^2 \, t_i \tag{2.3}$$

is the bending stiffness of this segment, E being Young's modulus.

The weight of the core is fixed by the given segment lengths ℓ_i and the values of b_i and h_i. The weight of the cover plates is proportional to $\sum_i b_i \ell_i t_i$ or, in view of (2.3), proportional to

$$\Omega = \sum_i \ell_i c_i / h_i^2 . \tag{2.4}$$

The virtual work C of the load P(x) on the deflection p(x) is called the compliance of the beam to this load :

$$C = \sum_i \int_{x_{i-1}}^{x_i} P p \, dx , \tag{2.5}$$

where the integration extends over the i-th segment, $x_{i-1} \leq x \leq x_i$, and the sum over all segments. The smaller the compliance, the stiffer is the structure. It therefore makes sense to design a structure for minimal weight when its compliance is prescribed. For this problem, a necessary and sufficient condition for global optimality will now be derived.

We first note that, by the principle of virtual work and equation (2.2), the compliance (2.5) may also be written in the form

$$C = \sum_i \int_{x_{i-1}}^{x_i} Q q \, dx = \sum_i c_i \ell_i \hat{q}_i^2 , \tag{2.6}$$

where

$$\hat{q}_i^2 = (1/\ell_i) \int_{x_{i-1}}^{x_i} q^2 \, dx \tag{2.7}$$

is the mean square curvature of the i-th segment.

If the designs c_i and c_i^* have the required compliance C and assume the curvatures q(x) and q*(x) under the load P(x) , we have

$$C = \sum_i c_i \ell_i \hat{q}_i^2 = \sum_i c_i^* \ell_i \hat{q}_i^{*2}. \tag{2.8}$$

For the considered problem, the virtual work of the loads has the prescribed value C. The principle of minimum potential energy thus becomes a principle of minimum strain energy. Applied to the design c_i^* , this principle yields the inequality

$$\sum_i c_i^* \ell_i \hat{q}_i^{*2} \leq \sum_i c_i^* \ell_i \hat{q}_i^2 , \qquad (2.9)$$

because the mean square curvatures $\hat{q}_i^2$ are computed from the curvature $q(x)$, which is kinematically admissible for the design c_i^* . Substituting (2.9) into (2.8), we obtain the inequality

$$\sum_i (c_i^* - c_i) \ell_i \hat{q}_i^2 \geq 0 . \qquad (2.10)$$

The differences $c_i^* - c_i$ appearing in (2.10) are subject to further inequalities. Because a bending stiffness is nonnegative, we have

$$(c_i^* - c_i) \ell_i \geq 0 \quad \text{if } t_i = \bar{t} , \text{ and} \qquad (2.11)$$

$$-(c_i^* - c_i) \ell_i \geq 0 \quad \text{if } t_i = \bar{\bar{t}} . \qquad (2.12)$$

Finally, if the design c_i is to have minimal weight, we must have $\Omega^* - \Omega \geq 0$ and hence

$$\sum_i (c_i^* - c_i) \ell_i / h_i^2 \geq 0 , \qquad (2.13)$$

by (2.4).

According to a theorem of Farkas[1], the inequality (2.13) follows from the inequalities (2.10), (2.11), and (2.12) if and only if it is a nonnegative linear combination of the latter. If the coefficients of this linear combination are denoted by $1/k^2$, μ_i/k^2 , and $\bar{\mu}_i/k^2$, with $\mu_i = \bar{\mu}_i = 0$ when $\bar{t} < t_i < \bar{\bar{t}}$, Farkas' theorem furnishes the necessary and sufficient optimality condition

$$h_i^2 \, \hat{q}_i^2 \begin{cases} = k^2 & \text{if } \bar{t} < t_i < \bar{\bar{t}} \,, \\[2mm] \leq k^2 & \text{if } t_i = \bar{t} \,, \\[2mm] \geq k^2 & \text{if } t_i = \bar{\bar{t}} \,. \end{cases} \tag{2.14}$$

Since it was not assumed that c_i and c_i^* are neighboring designs, this optimality condition has <u>global</u> character. When the constraints (2.1) are
not effective, and h_i is independent of i, the optimality condition (2.14)
assumes the form $\hat{q}_i^2$ = const, which was given by Sheu and Prager[2].

For the application of the optimality condition (2.14) it is useful to
recognize that, in view of (2.2), the mean square curvature (2.7) of the
i-th segment may be written in the form

$$\hat{q}_i^2 = (c_i \, \ell_i \,)^{-1} \int_{x_{i-1}}^{x_i} Q \, q \, dx \,, \tag{2.15}$$

where the integral may be evaluated according to (1.6).

<u>2.2. Example</u>. To illustrate the use of the optimality condition (2.14),
consider the continuous beam in Fig. 2.1a, which is loaded at the right
by a terminal couple P. The two spans are to have the same core dimensions but possibly different cover plate thicknesses t_i , (i = 1, 2), that
are subject to the constraint (2.1). The corresponding constraint on the
bending stiffnesses will be written as

$$\bar{c} \leq c_i \leq \bar{\bar{c}} \,. \tag{2.16}$$

If neither c_1 nor c_2 assume the values $\bar{c}$ or $\bar{\bar{c}}$, the optimality condition
(2.14) requires that

$$\hat{q}_1^2 = \hat{q}_2^2 \,. \tag{2.17}$$

Figure 2.1b shows the individual spans with their terminal moments. Setting $\ell_i / c_i = \ell_i'$, we have the following expressions for the terminal ro-

tations:

$$\theta_1 = R\, \ell_1' / 3 = (P - 2 R)\, \ell_2' / 6 \, ,$$

$$\theta_2 = (2 P - R)\, \ell_2' / 6 \, .$$

(2.18)

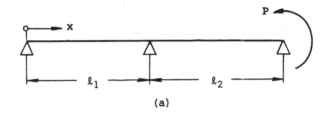

(a)

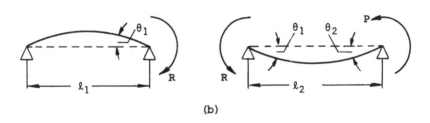

(b)

Fig. 2.1: Optimal design of continuous sandwich beam
for given compliance to terminal couple P

It follows from the first line of (2.18) that

$$R = P\, \ell_2' / 2\, (\ell_1' + \ell_2') \, .$$

(2.19)

Substitution of this value of R into the second line of (2.18) yields the
following expression for the compliance:

$$C = P\, \theta_2 = P^2\, \ell_2'\, (4\, \ell_1' + 3\, \ell_2') / 12\, (\ell_1' + \ell_2') \, .$$

(2.20)

We shall denote by $\bar{C}$ and $\bar{\bar{C}}$ the values obtained from (2.20) when c_1 = $c_2 = \bar{c}$ or $c_1 = c_2 = \bar{\bar{c}}$. Note that $\bar{C} > \bar{\bar{C}}$.

According to the remark made following equation (2.15),

$$\hat{q}_1^2 = R\,\theta_1 / c_1\,\ell_1 \;, \quad \hat{q}_2^2 = -\,(R\,\theta_1 - P\,\theta_2\,)\,/\,c_2\,\ell_2 \qquad (2.21)$$

With the use of (2.19) through (2.21), the optimality condition (2.17) furnishes a quadratic equation for $\rho = c_1/c_2$, whose relevant root is

$$\rho = \sqrt{(\,1 - \lambda^2\,)\,/\,3} - \lambda\;, \quad (\lambda = \ell_1\,/\,\ell_2\,)\;. \qquad (2.22)$$

Since ρ is nonnegative, it follows from (2.22) that the preceding analysis implies $\lambda \leq 1/2$. Note that (2.22) then yields values of ρ that range from 0 for $\lambda = 1/2$ to $\sqrt{1/3}$ for $\lambda = 0$.

In the following, we shall assume that the prescribed value of C satisfies the relation $\bar{\bar{C}} \leq C \leq \bar{C}$ as otherwise the problem does not admit a solution. The optimal design is then found as follows.

(1) $\bar{c} = 0$:

    (a) $\lambda < 1/2$: determine ρ from (2.22) and c_2 from (2.20), which may be written as

$$c_2 = (P^2\ell_2/12\,C)\,(4\lambda+3\rho)\,/\,(\lambda+\rho)\;, \qquad (2.23)$$

    and c_1 from $c_1 = \rho c_2$. If $c_2 > \bar{\bar{c}}$, go to (3).

    (b) $\lambda \geq 1/2$: the optimal design may not have the first span: $c_1 = 0$, $c_2 = P^2\ell_2/3C$. If $c_2 > \bar{\bar{c}}$, go to (3).

(2) $\bar{c} = 0$:

    (a) $\lambda < 1/2$: determine c_1 and c_2 as in (1a). If $c_1 < \bar{c}$, go to (2b).

    (b) $\lambda \geq 1/2$: set $c_1 = \bar{c}$ and determine c_2 from (2.23), which may now be written as

$$(P^2\ell_2/4\bar{c}c)\,\rho^2 - (1-\lambda P^2\ell_2/3\bar{c}c)\,\rho - \lambda = 0\;. \qquad (2.24)$$

If $c_2 > \bar{\bar{c}}$, go to (3).

(3) Set $c_2 = \bar{\bar{c}}$ and determine c_1 from (2.23), which may now be written as

$$\rho = \lambda\{ (P^2 \ell_2/3C) - \bar{\bar{c}}\}/(\bar{\bar{c}} - P^2 \ell_2/4C) . \tag{2.25}$$

It is worth noting how complex the "straightforward" optimization becomes even for such a simple structure when upper and lower bounds are prescribed for the bending stiffnesses.

In the limiting case, where the given core height 2 h and the unknown plate thickness t are allowed to vary continuously, the expression $h_i^2 \, \hat{q}_i^2$ on the left of the optimality condition (2.14) must be replaced by $h^2(x) \, q^2(x)$. For the example in Fig. 2.1, where h = const, this means that, under the action of the terminal couple P, the optimal beam will have a curvature q(x) satisfying

$$\left| q(x) \right| = k/h \tag{2.26}$$

if $\bar{c} = 0$ and if everywhere $c(x) < \bar{\bar{c}}$.

Assuming first that $\ell_1 \leq \ell_2$, and setting $L = \ell_1 + \ell_2$, we determine the distance $\xi \geq \ell_1$ between the leftmost support and the cross section at which the curvature (2.26) must change sign if the deflection is to vanish at the three supports. One readily finds

$$\xi = L - \sqrt{L \ell_2 / 2} . \tag{2.27}$$

Since the bending moment Q(x) must also change sign at $x = \xi$, we have

$$Q(x) = P (x - \xi) / (L - \xi) \quad \text{for } \ell_1 \leq x \leq L \tag{2.28}$$

and hence

$$Q(x) = - P x (\xi - \ell_1) / (L - \xi) \quad \text{for } 0 \leq x \leq \ell_1 . \tag{2.29}$$

In view of the optimality condition (2.26), the bending stiffness of the optimal beam equals

$$c = h \, |Q| \, / \, k \, , \tag{2.30}$$

where k must be chosen in such a manner that the beam has the prescribed compliance to the couple P.

Note that for $\ell_1 / \ell_2 \to 0$, our analysis covers the beam that is built in at $x = 0$ and simply supported at $x = \ell_2$.

2.3. Sandwich beams subject to alternative compliance constraints. In general, a structure has to be designed for several alternative states of loading. It may, of course, happen that only one of these is <u>relevant</u> in the sense that it influences the choice of structural dimensions, but this case is comparatively rare. In this section, we shall therefore discuss the optimal design of a sandwich beam of the type considered at the beginning of Section 2.1 that is subject to two alternative states of loading with generalized loads $P_{\alpha 1}$ and $P_{\alpha 2}$. The beam is to be designed for minimum weight subject to the condition that its compliances C_1 and C_2 to the two states of loading satisfy the inequalities

$$C_1 \leq C' \, , \quad C_2 \leq C'' \tag{2.31}$$

for given values of C' and C" .

If only one of the two states of loading is relevant, the problem reduces to that treated in Section 2.1. We shall therefore assume that both states of loading are relevant, that is, that $C_1 = C'$ and $C_2 = C''$. Inequalities of the form (2.10) then hold for both states of loading. If $\hat{q}_{i1}^2$ and $\hat{q}_{i2}^2$ are the mean square curvatures of the i-th segment for the two states of loading, we have

$$\sum_i \, (\, c_i^\star - c_i \,) \, \ell_i \, \hat{q}_{i1}^2 \geq 0 \, , \tag{2.32}$$

$$\sum_i (c_i^* - c_i) \, \ell_i \, \hat{q}_{i2}^2 \geq 0 , \tag{2.33}$$

instead of (2.10), while (2.11), (2.12), and (2.13) remain valid. Use of Farkas' theorem with the nonnegative multipliers λ_1^2, λ_2^2, μ_i, and $\bar{\mu}_i$ for the inequalities (2.32), (2.33), (2.11), and (2.12) now furnishes the optimality condition

$$h_i^2 (\lambda_1^2 \, \hat{q}_{i1}^2 + \lambda_2^2 \, \hat{q}_{i2}^2) \begin{cases} = 1 & \text{if } \bar{t} < t_i < \bar{\bar{t}} , \\[4pt] \leq 1 & \text{if } t_i = \bar{t} , \\[4pt] \geq 1 & \text{if } t_i = \bar{\bar{t}} . \end{cases} \tag{2.34}$$

A beam is optimal if and only if there exist nonnegative multipliers λ_1^2 and λ_2^2 such that its mean square curvatures under the two states of loading satisfy (2.34). The generalisation of the optimality condition (2.34) to more than two states of loading is obvious.

In the limiting case, where the given core height 2 h and the unknown plate thickness t are allowed to vary continuously, the left side of (2.34) must be replaced by $h^2(x) \{ \lambda_1^2 \, q_1^2(x) + \lambda_2^2 \, q_2^2(x) \}$, where $q_1(x)$ and $q_2(x)$ are the curvatures of the optimal beam under the two states of loading.

For brevity, the following discussion of the use of the optimality condition (2.34) will be restricted to a statically determinate beam for which the constraints (2.1) are not active. With the mean square bending moments

$$\hat{Q}_{i\gamma}^2 = \ell_i^{-1} \int_{x_{i-1}}^{x_i} Q_\gamma^2 \, dx , \qquad (\gamma = 1, 2) , \tag{2.35}$$

the optimality condition (2.34) then has the form

$$c_i = \lambda_1 \, h_i \, R_i , \tag{2.36}$$

where

$$R_i = (\hat{Q}_{i1}^2 + \beta^2 \, \hat{Q}_{i2}^2)^{1/2} , \qquad \beta = \lambda_2 / \lambda_1 . \tag{2.37}$$

The compliances of the beam to the two loadings are

$$c_\gamma = \sum_i \hat{Q}_{i\gamma}^2 \ell_i / c_i , \quad (\gamma = 1, 2) . \tag{2.38}$$

We substitute (2.36) into (2.38), solve for λ_1 , and use the resulting expression in (2.36). Thus,

$$c_i = (h_i R_i / c_\gamma) \sum_i \hat{Q}_{i\gamma}^2 \ell_i / h_i R_i , \quad (\gamma = 1, 2) . \tag{2.39}$$

It follows from the two equations (2.39) that

$$\rho(\beta) = C_2 / C_1 = N / D , \text{ where}$$

$$N = \sum_i \hat{Q}_{i2}^2 \ell_i / h_i (\hat{Q}_{i1}^2 + \beta^2 \hat{Q}_{i2}^2)^{1/2} , \tag{2.40}$$

$$D = \sum_i \hat{Q}_{i1}^2 \ell_i / h_i (\hat{Q}_{i1}^2 + \beta^2 \hat{Q}_{i2}^2)^{1/2} .$$

If $C_1 = C'$ but $C_2 < C''$, we have $\lambda_2 = 0, \beta = 0$, and hence

$$\rho(0) = (\sum_i \hat{Q}_{i2}^2 \ell_i / h_i \hat{Q}_{i1}^2) / (\sum_i \hat{Q}_{i1}^2 \ell_i / h_j) < C'' / C' . \tag{2.41}$$

Similarly, if $C_2 = C''$ but $C_1 < C'$, we have $\lambda_1 = 0$, $\beta = \infty$ and hence

$$\rho(\infty) = (\sum_i \hat{Q}_{i2}^2 \ell_i / h_i) / (\sum_i \hat{Q}_{i1}^2 \ell_i / h_i \hat{Q}_{i2}^2) > C'' / C' . \tag{2.42}$$

Accordingly both compliance constraints are active if

$$\rho(0) \leq C'' / C' \leq \rho(\infty) . \tag{2.43}$$

If the condition (2.43) is satisfied, the optimal design may be found as follows. Since the given value of $\rho = C'' / C'$ exceeds $\rho(0)$, we compute $\rho(\beta)$ from (2.40) for increasing values of β until $\rho(\beta) > C'' / C'$, and then determine by interpolation the value of β for which $\rho(\beta) = C'' / C'$. The optimal bending stiffnesses are then found from equation (2.39), in which either $\gamma = 1$ or $\gamma = 2$ may be used.

Note that the relations (2.36) through (2.42) remain valid if the beam is statically indeterminate, but the mean square bending moments now contain terms in the redundants. For each redundant X_k , an additional equation is furnished by the condition that the true curvature $Q \, / \, c$ must be orthogonal to the bending moment Q_k caused by $X_k = 1$, when all other redundants vanish and the beam does not carry any loads:

$$\sum_i c_i^{-1} \int_{x_{i-1}}^{x_i} Q \, Q_k = 0 \ . \tag{2.44}$$

For numerical examples, the reader is referred to papers by Chern and Prager[3] , Martin[4] , and Chern and Martin[5] .

3. Optimality Conditions for Other Structures and Constraints

3.1. Optimal design of other elastic structures subject to compliance constraints. For ease of exposition, only beams were treated in Chapter 2, but the method by which the various optimality conditions were derived applies equally well to other elastic structures subject to compliance constraints. The following examples will illustrate this remark.

(a) Sandwich member subject to axial force and bending moments. Consider a beam that is simply supported at $x = 0$ and $x = \ell$ and subject to both an axial tensile force P_1 and a uniformly distributed transverse load of the intensity P_2 per unit length. The compliance of the beam to these loads is to have the value C. The beam is to have sandwich construction with a uniform core of width b and height 2 h , and segmentwise constant thicknesses of cover plates, the segment boundaries being given. For the i-th segment, the possibly different thicknesses of the upper and lower plates will be denoted by t_i' and t_i'' . These thicknesses are to be chosen to minimize the total weight of cover plates, which is proportional to

$$\Omega = \sum_i (t_i' + t_i'') \, \ell_i \, , \tag{3.1}$$

where ℓ_i is the length of the i-th segment.

Denote the axial strains in the upper and lower cover plates by $q'(x)$ and $q''(x)$, and define the mean square strains in the i-th segment by

$$\hat{q}_i'^2 = \ell_i^{-1} \int_{x_{i-1}}^{x_i} q'^2 \, dx \, , \qquad \hat{q}_i''^2 = \ell_i^{-1} \int_{x_{i-1}}^{x_i} q''^2 \, dx \, . \tag{3.2}$$

The compliance may now be written as

$$C = E \, b \sum_i (t_i' \, \hat{q}_i'^2 + t_i'' \, \hat{q}_i''^2) \, \ell_i \, , \tag{3.3}$$

where E is Young's modulus of the plate material. Proceeding as in
Section 2.1, we consider a second design $t_i'^*$, $t_i''^*$ with the same compliance and use the principle of minimum potential energy to derive
the inequality

$$\sum_i \{ (t_i'^* - t_i') \hat{q}_i'^2 \ell_i + (t_i''^* - t_i'') \hat{q}_i''^2 \ell_i \} \geq 0 . \qquad (3.4)$$

Since the thicknesses of the cover plates are nonnegative, we have

$$t_i'^* - t_i' \geq 0 \quad \text{if } t_i' = 0 ,$$
$$\qquad\qquad\qquad\qquad\qquad\qquad\qquad (3.5)$$
$$t_i''^* - t_i'' \geq 0 \quad \text{if } t_i'' = 0 .$$

Finally, if the design t_i' , t_i'' is to be optimal, we must have

$$\sum_i \{ (t_i'^* - t_i') \ell_i + (t_i''^* - t_i'') \ell_i \} \geq 0 . \qquad (3.6)$$

Applying Farkas' theorem to the inequalities (3.4), (3.5), and (3.6),
we obtain the optimality conditions

$$\hat{q}_i'^2 \begin{cases} = k^2 & \text{if } t_i' > 0 , \\ \leq k^2 & \text{if } t_i' = 0 , \end{cases}$$

$$\qquad\qquad\qquad\qquad\qquad\qquad\qquad (3.7)$$

$$\hat{q}_i''^2 \begin{cases} = k^2 & \text{if } t_i'' > 0 , \\ \leq k^2 & \text{if } t_i'' = 0 . \end{cases}$$

We briefly discuss the application of these optimality conditions to
the limiting case where the thicknesses $t'(x)$ and $t''(x)$ of the cover
plates vary continuously. If $t'(x)$ and $t''(x)$ are positive all along
the span $0 \leq x \leq \ell$, the optimality conditions (3.7) furnish $q'(x) =$
$q''(x) = k$, and the axial force Q_1 and the bending moment Q_2 are given by

$$Q_1(x) \equiv P_1 = E b k (t' + t'') , \qquad\qquad (3.8')$$

$$Q_2(x) \equiv P_2 \, x \, (\ell - x) \, / \, 2 = E \, b \, h \, k \, (-t' + t'') \, . \qquad (3.8'')$$

Solving for t' and t'' , we find

$$\left.\begin{matrix} t' \\ \\ t'' \end{matrix}\right\} = \{2 \, P_1 \, h \mp P_2 \, x \, (\ell - x)\} \, / \, (4 \, E \, b \, h \, k) \, . \qquad (3.9)$$

Since t' is nonnegative, (3.9) shows that our analysis is valid only if

$$P_2 \leq 8 \, P_1 \, h \, / \, \ell^2 \, . \qquad (3.10)$$

In this case, the value of k in (3.9) may be determined from the prescribed compliance

$$C = E \, b \, k^2 \int_0^\ell (t' + t'') \, dx \, . \qquad (3.11)$$

Substituting (3.8) into (3.11), we find

$$k = C \, / \, (P_1 \, \ell) \, . \qquad (3.12)$$

If (3.10) is satisfied, the optimal design is therefore given by

$$\left.\begin{matrix} t' \\ \\ t'' \end{matrix}\right\} = P_1 \, \ell \, \{2 \, P_1 \, h \mp P_2 \, x \, (\ell - x)\} \, / \, (4 \, E \, b \, h \, C) \, . \quad (3.13)$$

We leave it to the reader to discuss the case where (3.10) is not satisfied.

(b) Truss of given layout. Consider an elastic truss of given layout that is to have a prescribed compliance C to a given loading. Length and cross-sectional area of the i-th bar will be denoted by ℓ_i and A_i . To avoid designs in which some bars of the given layout are omitted, we prescribe a lower bound $\bar{A}$ for A_i :

$$A_i \geq \bar{A} \, . \qquad (3.14)$$

The cross-sectional areas are to be chosen to minimize the total weight of the bars of the truss, which is proportional to

$$\Omega = \sum_i \ell_i A_i = \sum_i V_i \, , \tag{3.15}$$

where V_i is the volume of bar i.

If q_i is the axial strain caused in bar i by the given loads, the compliance of the truss to these loads is

$$c = \sum_i V_i \, q_i^2 \, . \tag{3.16}$$

Proceeding as under (a) above, one readily obtains the following necessary and sufficient condition for global optimality:

$$q_i^2 \begin{cases} = k^2 & \text{if } A_i > \bar{A} \, , \\ \leq k^2 & \text{if } A_i = \bar{A} \, . \end{cases} \tag{3.17}$$

Since the optimal design of trusses will be treated in greater detail in Chapter 5, we do not discuss the use of the optimality condition (3.17) here.

(c) Sandwich plate. The derivation of the optimality condition for a sandwich plate with the median plane z = 0, the given core thickness 2 h(x,y), the minimum thickness $\bar{t}$ of the identical cover plates, and prescribed compliance to given transverse loading follows the same lines as the derivation of (2.26). If p(x,y) is the deflection, and $q_1 = -\partial^2 p/\partial x^2$, $q_2 = -\partial^2 p/\partial y^2$, and $q_3 = -\partial^2 p/\partial x \, \partial y$ are the curvatures and the twist for the coordinate directions, the optimality condition has the form

$$(1-\nu)(q_1^2 + q_2^2 + 2q_3^2) + \nu(q_1+q_2)^2 \begin{cases} = k^2/h^2 & \text{if } t > \bar{t} \, , \\ \leq k^2/h^2 & \text{if } t = \bar{t} \, , \end{cases} \tag{3.18}$$

where ν is Poisson's ratio.

3.2. Optimal design for other constraints. As was pointed out by Prager
and Taylor[6], the procedure by which the optimality conditions (2.14) and
(2.34) were obtained may be used whenever each constraint concerns a quantity, such as compliance, that is characterized by a minimum principle,
such as the principle of minimum strain energy used above. The condition
obtained in this way will be necessary and sufficient for global optimality if the minimum characterization of each constrained quantity has
global character. The following examples illustrate these remarks.

(a) Design for given fundamental frequency. Consider a cantilever beam of
   length ℓ carrying a given nonstructural mass m at the tip $x = \ell$. The
   beam is to have segmentwise constant sandwich section of the kind
   specified in the beginning of Section 2.1; it is to be designed for
   minimum weight when its fundamental frequency ω is prescribed.

   For the i-th segment, the specific core mass will be written as $A\, h_i$,
   and the specific mass of the two cover plates will be written as
   $B\, h_i^{-2}\, c_i$ (see equation (2.3)), where A and B are independent of i.
   The mean square curvature of the i-th segment is defined by (2.7) and
   the mean square deflection by

   $$\hat{p}_i^2 = \ell_i^{-1} \int_{x_{i-1}}^{x_i} p^2 \, dx \; . \tag{3.19}$$

According to Rayleigh, we have

   $$\omega^2 = \min (N/D) \quad \text{with} \quad N = \sum_i c_i \, \ell_i \, \hat{q}_i^2 \; ,$$

   $$D = m\, p^2(\ell) + \sum_i (A\, h_i + B\, h_i^{-2}\, c_i) \, \ell_i \, \hat{p}_i^2 \; , \tag{3.20}$$

   where the mean square deflections $\hat{p}_i^2$ and curvatures $\hat{q}_i^2$ correspond to
   any kinematically admissible deflection $p(x)$.

   If c_i and c_i^* are designs with the fundamental frequency ω , and if the
   values $\hat{p}_i^2$ and $\hat{q}_i^2$ correspond to the fundamental mode of the design c_i ,
   it follows from the minimum characterization of ω^2 that

$$\omega^2 \leq N^*/D^* \quad \text{with } N^* = \sum_i c_i^* \, \ell_i \, \hat{q}_i^2 \, , \tag{3.21}$$

$$D^* = m \, p^2(\ell) + \sum_i (A \, h_i + B \, h_i^{-2} \, c_i^*) \, \ell_i \, \hat{p}_i^2 \, ,$$

while the indication "min" may be dropped from (3.20). Both sides of this form of (3.20) are now multiplied by the denominator D and collected on the right; (3.21) is transformed in the same manner, and the equation is subtracted from the inequality. Thus,

$$\sum_i (c_i^* - c_i) \, \ell_i \, \{ \hat{q}_i^2 - \omega^2 \, B \, h_i^{-2} \, \hat{p}_i^2 \} \geq 0 \, . \tag{3.22}$$

Application of Farkas' theorem to the inequalities (3.22), (2.11), (2.12), and (2.13) finally furnishes the following necessary and sufficient condition for global optimality of the design c_i :

$$h_i^2 \, \hat{q}_i^2 - \omega^2 \, B \, \hat{p}_i^2 \begin{cases} = k^2 & \text{if } \bar{t} < t_i < \bar{\bar{t}} \, , \\ \leq k^2 & \text{if } t_i = \bar{t}_i \, , \\ \geq k^2 & \text{if } t_i = \bar{\bar{t}}_i \, . \end{cases} \tag{3.23}$$

For examples of optimal design with frequency constraint, the reader is referred to papers by Niordson[7], Turner[8], Taylor[9], Sheu[10], McCart Haug, and Streeter[11], Zarghamee[12], and Karihaloo and Niordson[13].

(b) Design for given load factor for plastic collapse. Consider a rigid, perfectly plastic truss of given layout that is subject to a given loading. Length and cross-sectional area of the i-th bar will be denoted by ℓ_i and A_i . The cross-sectional areas are to satisfy the inequality (3.14), and the weight of the truss, or what amounts to the same, the quantity Ω in (3.15) is to be minimized subject to the constraint that the load factor for plastic collapse is to have a given value λ . Let $\pm\sigma_0$ be the tensile and compressive yield limits of the material used for the bars of the truss, and denote by q_i the axial strain rate of bar i in any normalized collapse mechanism, that is, a mechanism for which the power of the given loads is unity. The inter-

nal power of dissipation for the i-th bar in this mechanism then is
$\sigma_0 \, |q_i| \, V_i$, where $V_i = \ell_i \, A_i$ is the volume of this bar. The kinematic theorem of limit analysis thus furnishes the following minimum
characterization of the load factor for plastic collapse:

$$\lambda = \sigma_0 \, \min \sum_i |q_i| \, V_i \; . \tag{3.24}$$

If the designs V_i and V_i^* have the required load factor, and q_i is
the axial strain rate of bar i in the normalized collapse mechanism
of the design V_i under the given loads, it follows from (3.24) that

$$\sum_i (V_i^* - V_i) \, |q_i| \geq 0 \; . \tag{3.25}$$

Applying Farkas' theorem to (3.25) and the inequalities

$$V_i^* - V_i \geq 0 \quad \text{if} \quad V_i = \bar{V}_i = \ell_i \, \bar{A} \, , \tag{3.26}$$

$$\sum_i (V_i^* - V_i) \geq 0 \quad \text{if the design } V_i \text{ is optimal,} \tag{3.27}$$

we obtain the optimality condition

$$| \, q_i \, | \begin{cases} = k \quad \text{if} \quad V_i > \bar{V}_i \; , \\[2mm] \leq k \quad \text{if} \quad V_i = \bar{V}_i \; , \end{cases} \tag{3.28}$$

which is necessary and sufficient for global optimality of the design
V_i . Since the time scale of collapse is not relevant, the constant k
in (3.28) may be chosen as unity. When both sides of the resulting
condition are multiplied by V_i , it is seen that the optimal design
admits a collapse mechanism in which the contribution of any bar to
the internal power of dissipation of the truss is numerically equal
to or less than its contribution to the weight of the truss, depending
on whether the cross section of the considered bar is greater than or
equal to $\bar{A}$. Except for the consideration of a lower bound on cross-
sectional area, this form of the optimality condition is due to Drucker and Shield[14]. Optimal plastic design of trusses will be conside-

red in Chapter 5.

Constraints that have been handled in this manner include: elastic buck-
ling load (Taylor[15], Taylor and Liu[16]), rate of compliance in steady
creep (Prager[17]), dynamic elastic compliance under harmonically varying
loads (Icerman[18], Mróz[19], Plaut[20]), and elastic deflection at a specified point (Shield and Prager[21], Chern and Prager[22], Chern[23], Prager[24]).
For the first two types of constraint, classical minimum principles could
be used, and for the third type, a suitable minimum principle was derived
in the first-named paper. For the fourth type of constraint, a principle
of stationary mutual potential energy was established in the first-named
paper, which yields a global optimality condition only for a statically
determinate structure. For redundant structures, however, this principle
only furnishes a condition for the structural weight to be stationary in
the neighborhood of the considered design.

3.3. A three-dimensional problem. This section is concerned with a fairly
general problem of structural optimization. A three-dimensional body B
that consists of a given material is to be designed for minimum weight
subject to the following constraints.

(a) Geometrical constraints. The surface S of B is to contain a given part
    S_1 , which is rigidly supported, and a given part S_2 , which is subjec-
    ted to given surface tractions. The remainder S_3 of S, which is to be
    free from surface tractions, is at the choice of the designer subject
    to the condition that it must remain inside the region V_0 that has the
    given surface S_0 .

(b) Behavioral constraint. A scalar quantity (e.g. the elastic complian-
    ce) that represents a relevant feature of the mechanical behavior of
    B is to have a given value C, which is supposed to be characterized
    by a global minimum principle of the form

$$C = \min \int F\{\bar{r}\}\, dV \,.\tag{3.29}$$

Here, $\bar{r} = \bar{r}(x)$ is a certain field, for instance a stress field, which must be admissible in the sense that it must satisfy certain differential equations, and boundary and continuity conditions. $F\{\bar{r}\}$ is a positive definite functional of $\bar{r}$, and the integration is extended over the volume V of B. The minimum in (3.29) is attained when $\bar{r} = r$, where r is the actual field set up in B by the given surface tractions on S_2. For example, when C is the elastic compliance of B, then $\bar{r}$ is an arbitrary kinematically admissible strain field, and $F\{\bar{r}\}$ is the corresponding specific strain energy.

Let B* be a second body that satisfies the same geometrical and behavioral constraints as B, but has a different traction-free surface S_3^*, and denote by V* the region occupied by B*. The surface S_3^* will, in general, be partly inside and partly outside of V, and the region V* may be obtained from V by adding the region V^+ bounded by S_3 and the exterior part of S_3^*, and subtracting the region V^- bounded by S_3 and the interior part of S_3^*.

Since B and B* satisfy the behavioral constraint,

$$C = \int F\{r\}\, dV = \int F\{r*\}\, dV* \,,\tag{3.30}$$

where r* is the actual field in B*. If the actual field r in B can be continued in an admissible manner throughout the region V_0, the continued field is admissible for B*. It then follows from the minimum characterization (3.29) that

$$\int F\{r*\}\, dV* \le \int F\{r\}\, dV* = \int F\{r\}\, dV + \int F\{r\}\, dV^+ - \int F\{r\}\, dV^- \,.\tag{3.31}$$

Substitution of (3.31) into (3.30) furnishes

$$\int F\{r\}\, dV^+ - \int F\{r\}\, dV^- \ge 0 \,.\tag{3.32}$$

In general, S_3 will include parts S_3' that lie on S_0 . If the remainder of S_3 is denoted by S_3'' , the following conditions are sufficient for global optimality of the design B :

$$F = F^0 = const \quad on \ S_3'' \ with \ F^0 > 0 \ ,$$

$$F = F^0 - F^+ \quad in \ v^+ \ with \ F^+ \geq 0 \ , \tag{3.33}$$

$$F = F \ + F^- \quad in \ v^- \ with \ F^- \geq 0 \ .$$

To prove this, we substitute (3.33) into (3.32) to obtain

$$F^0 \ (\ v^+ - v^- \ ) \geq \int F^+ \ dv^+ + \int F^- \ dv^- \geq 0 \tag{3.34}$$

because the functional F is positive definite. Since $v^+ - v^-$ on the left side of (3.34) is the excess of the volume of the design B* over that of the design B, and since $F^0 \geq 0$, it follows from (3.34) that the design B* cannot be lighter than the design B .

Extension of the optimality conditions (3.33) to the case of several behavioral constraints of the form considered above follows the steps that led to the conditions (2.34) and will not be discussed here.

Problems of the kind considered in this section were first discussed by Mroz[25] with reference to optimal plastic design, and more generally by Prager[26,27]. More recently, optimal design for given dynamic compliance to harmonically varying loads (Mroz[28]) and optimal plastic design of disks (Kozlowski and Mroz[29]) were discussed in these terms. For examples, the reader is referred to these papers.

PART II

OPTIMIZATION OF STRUCTURAL LAYOUT

4. A General Theory of Optimal Plastic Design - Optimal Division
into Elements of Prescribed Shape

4.1. A general theory of optimal plastic design. Since most problems
treated in Part II concern plastic design, it will be useful to review a
general theory of optimal plastic design, which was initiated by Marçal
and Prager[30], and Prager and Shield[31], and extended by Charrett and Rozvany[32], and Save[33].

Consider a one- or two-dimensional rigid, perfectly plastic structure oc-
cupying a region V, which consists of a number of subregions V_i . At the
point x of the subregion V_i , let the structural design be specified by a
number of design parameters $t_{ij}(x)$ that vary with x according to

$$t_{ij}(x) = \sum_k \tau_{ijk} \phi_{ijk}(x) \quad , \quad (\text{ no summation over i, j }) \qquad (4.1)$$

where τ_{ijk} is independent of x and ϕ_{ijk} is a given function of x (shape
function). For example, if the thickness t_{i1} of the upper cover plate of
a sandwich beam with uniform rectangular core section of given height and
width is to vary linearly between $x = x_{i-1}$ and $x = x_i = x_{i-1} + \ell_i$, we have

$$t_{i1}(x) = \tau_{i11} (x_i - x) / \ell_i + \tau_{i12} (x - x_{i-1}) / \ell_i \qquad (4.2)$$

where the thicknesses $t_{i1}(x_{i-1}) = \tau_{i11}$ and $t_{i1}(x_i) = \tau_{i12}$ are at the choice of the designer, but the shape functions $\phi_{i11} = (x_i - x) / \ell_i$ and
$\phi_{i12} = (x - x_{i-1}) / \ell_i$ are prescribed.

It will be assumed in the following that the design parameters are chosen
in such a manner that the specific rate of dissipation at point i of sub-

region V_i may be written as

$$D_i(x;q) = \sum_j \sum_k \tau_{ijk} \, \phi_{ijk}(x) \, d_{ijk}(q) \, , \qquad (4.3)$$

where $q = q(x)$ is the strain rate vector at x and the dissipation functions d_{ijk} , which depend on q but are independent of the design parameters τ_{ijk} , are homogeneous of the order one and convex (see (1.29) and
(1.30)). For example, since the upper cover plate of the sandwich beam
considered above makes the contribution b h $t_{i1}(x) \, |q|$ to $D_i(x;q)$, where
b and 2 h are the width and height of the core and q is the rate of curvature at x, we have

$$d_{i11} = d_{i12} = b \, h \, |q| \, . \qquad (4.4)$$

The objective function Ω , which is to be minimized, and which need not be
the structural weight, will be called cost; it will be written in the form

$$\Omega = \sum_i \int_{V_i} \omega_i(t) \, dx \, , \qquad (4.5)$$

where $t = t(x)$ is the design vector with components $t_{ij}(x)$ and the specific cost ω_i in subregion V_i is supposed to be a convex function of these
components. For any two design vectors $t* = t*(x)$ and $t = t(x)$, we therefore have

$$\omega_i(t*) - \omega_i(t) \geq \sum_j \, (\, t^*_{ij} - t_{ij} \,) \, (\, \partial\omega_i/\partial t_{ij} \,)_t$$

$$= \sum_j \sum_k \, (\, \tau^*_{ijk} - \tau_{ijk} \,) \, \phi_{ijk} \, (\, \partial\omega_i/\partial t_{ij} \,)_t \quad (4.6)$$

where the subscript t indicates that the derivative is to be evaluated for
the design t.

The optimization problem, which we propose to investigate, may be stated
as follows. The design parameters τ_{ijk} are to be chosen to minimize the
objective function Ω subject to the constraints

$$\bar{\tau}_{ijk} \leq \tau_{ijk} \leq \bar{\bar{\tau}}_{ijk} \, , \qquad (4.7)$$

where $\bar{\tau}_{ijk}$ and $\bar{\bar{\tau}}_{ijk}$ are given, and the further constraint that none of a given set of loadings should exceed the load carrying capacity of the structure.

Discussing first the case of a single loading, we consider the designs τ_{ijk} and τ^*_{ijk} , the first of which is at collapse under the given loading whereas the second is at or below collapse. It follows from the kinematic theorem of limit analysis that, for the collapse mechanism $q = q(x)$ of the first design, this design has an internal power of dissipation that cannot be smaller than that of the second design. Thus,

$$\sum_i \sum_j \sum_k (\tau^*_{ijk} - \tau_{ijk}) \int_{V_i} \phi_{ijk}(x) \, d_{ijk}(q(x)) \, dx \geq 0. \qquad (4.8)$$

We now apply Farkas' theorem to (4.8), the inequalities

$$\tau^*_{ijk} - \tau_{ijk} \geq 0 \quad \text{if } \tau_{ijk} = \bar{\tau}_{ijk} \, ,$$

$$-(\tau^*_{ijk} - \tau_{ijk}) \geq 0 \quad \text{if } \tau_{ijk} = \bar{\bar{\tau}}_{ijk} \, , \qquad (4.9)$$

and the inequality

$$\sum_i \sum_j \sum_k (\tau^*_{ijk} - \tau_{ijk}) \int_{V_i} \phi_{ijk}(x) \, (\partial\omega_i/\partial t_{ij})_t \, dx \geq 0 \qquad (4.10)$$

which, according to (4.6),excludes the possibility that the design τ^*_{ijk} may have a smaller cost than the design τ_{ijk} . We thus obtain the following sufficient condition for global optimality of the design τ_{ijk} :

$$\int_{V_i} \phi_{ijk}(x) \{ d_{ijk}(q(x)) - (\partial\omega_i/\partial t_{ij})_t \} dx \begin{cases} = 0 \text{ if } \bar{\tau}_{ijk} < \tau_{ijk} < \bar{\bar{\tau}}_{ijk} \, , \\ \leq 0 \text{ if } \tau_{ijk} = \bar{\tau}_{ijk} \, , \\ \geq 0 \text{ if } \tau_{ijk} = \bar{\bar{\tau}}_{ijk} \, . \end{cases}$$

$$(4.11)$$

In (4.11), it was not necessary to apply a positive factor λ to the term $d_{ijk}(q(x))$ because $\lambda \, d_{ijk}(q(x)) = d_{ijk}(\lambda q(x))$ by (1.29), and $\lambda q(x)$ is a collapse mechanism for the given loading if q(x) is such a mechanism.

Note that there is an optimality condition of the form (4.11) for each design parameter τ_{ijk} . To interpret these optimality conditions, we remark that the integrals

$$\int_{V_i} \phi_{ijk}(x)\,(\partial \dot{\omega}_i/\partial t_{ij})_t\,dx \quad \text{and} \quad \int_{V_i} \phi_{ijk}(x)\,d_{ijk}(q(x))\,dx \qquad (4.12)$$

represent the marginal rates of increase of cost and internal power of dissipation caused by an increase of the design parameter τ_{ijk} . The design τ_{ijk} is optimal if it admits a collapse mechanism for the given loading such that, for each design parameter τ_{ijk} , the marginal rate of increase of the internal power of dissipation is equal to, or not greater than, or not smaller than, the marginal rate of increase of the cost, depending on whether τ_{ijk} has a value between the bounds (4.7), or equals the lower or the upper bound.

It is important to note that, to qualify for use in the application of the optimality theorem, the collapse mechanism q(x) must correspond to a stress field Q(x) that is statically admissible for the given loading and nowhere exceeds the yield limit. According to a theorem of Horne[34], the given loading then represents the load carrying capacity of the design τ_{ijk}.

The condition (4.11), which is sufficient for global optimality, is also necessary, because it can be shown that the inequalities (4.8) and (4.10) become equations when τ_{ijk} and τ^*_{ijk} are neighboring designs.

The case of two alternative loadings, neither one of which is to exceed the load carrying capacity of the structure, can be treated along the lines of Section 2.3. The optimality condition that is obtained in this manner has the form of (4.11) except that the term $d_{ijk}(q(x))$ is replaced by

$$d_{ijk}(q'(x)) + d_{ijk}(q''(x)) \,, \qquad (4.13)$$

where q'(x) and q"(x) are the strain rate vectors of collapse mechanisms
under the two loadings. Contrary to what was the case in Section 2.3, it
is now not necessary to use multipliers, say λ' and $\lambda"$, for the two
terms of (4.13), because the fact that the dissipation functions d_{ijk}
are homogeneous of the order one (see (1.29)) enables us to absorb
these multipliers in the definitions of the collapse mechanisms. If one
of the loadings is not relevant, the structure does not deform under it,
and the corresponding strain rate vector and dissipation function vanish
identically. The optimality condition then takes the form (4.11).

The generalisation of the preceding discussion to more than two alternative loadings is straightforward. Of particular interest is the case of
a family of loadings that depends on a parameter ξ , which varies continuously from $\xi = \xi_0$ to $\xi = \xi_1 > \xi_0$. We then have a family of collapse
mechanisms $q(x;\xi)$, and the term $d_{ijk}(q(x))$ in the optimality condition
(4.11) must be replaced by

$$\int_{\xi_0}^{\xi_1} d_{ijk}(q(x;\xi)) \, d\xi \; . \tag{4.14}$$

For a beam, for instance, that is supported at x = 0 and x = ℓ,

$$P(x;\xi) = \begin{cases} P_0 & \text{for } \xi < x < \xi + h \\ 0 & \text{for } \xi + h < x < \ell \end{cases} \qquad (\, 0 \le \xi \le \ell - h \,) \tag{4.15}$$

represents a uniformly loaded segment of length h that may assume any position on the beam (moving load).

4.2. Examples. As first example, we consider a beam of rectangular cross
section of uniform height and linearly varying width that is simply supported at x = 0 and built in at x = ℓ , and carries a uniformly distributed load of the intensity P. As design parameters, we choose the yield
moments Y_1 and Y_2 at x = 0 and x = ℓ . With

$$\xi = x \, / \, \ell \, , \tag{4.16}$$

the variation of the yield moment along the beam is then given by

$$Y(\xi) = (1 - \xi) \, Y_1 + \xi \, Y_2 \; . \tag{4.17}$$

If the unknown reaction at $\xi = 0$ is written as $\rho \, P \, \ell$, the bending moment Q varies according to

$$Q(\xi) = P \, \ell^2 \, \xi \, (2 \rho - \xi) \, / \, 2 \; . \tag{4.18}$$

For yield hinges to occur at $\xi = \xi_0$ and $\xi = 1$, we must have

$$Y(\xi_0) = Q(\xi_0) \quad , \quad Y_2 = -Q(1) \; . \tag{4.19}$$

Moreover, the bending moment $Q(\xi)$ must not exceeed the yield moment $Y(\xi)$ in the neighborhood of $\xi = \xi_0$, and this requires that

$$Q'(\xi_0) = Y'(\xi_0) \; , \tag{4.20}$$

where the primes denote differentiation with respect to ξ .

To determine the value of ξ_0 , we note that, if the yield mechanism has the rate of rotation θ at $\xi = \xi_0$, it will have the rate of rotation $-\xi_0 \, \theta$ at $\xi = 1$. The internal power of dissipation then is

$$D = \{ \, Y(\xi_0) + \xi_0 \, Y_2 \, \} \, \theta \; . \tag{4.21}$$

With the use of (4.17), equation (4.21) may be written as

$$D = \{ \, Y_1 \, (1 - \xi_0) + 2 \, Y_2 \, \xi_0 \, \} \, \theta \; . \tag{4.22}$$

On the other hand, the weight of the beam is proportional to

$$\Omega = Y_1 + Y_2 \; . \tag{4.23}$$

Optimality requires that the derivatives of D with respect to Y_1 and Y_2 have the same ratio to each other as the derivatives of Ω . Thus,

$1 - \xi_0 = 2 \, \xi_0$, or

$$\xi_0 = 1/3 \ . \tag{4.24}$$

Using this value of ξ_0 and (4.18) through (4.20), we finally obtain

$$\rho = 7/18 \ , \qquad Y_1 = P\ell^2/18 \ , \qquad Y_2 = P\ell^2/9 \ . \tag{4.25}$$

As second example, we consider a rectangular opening ABCD (Fig. 4.1) that is spanned by the simply supported horizontal beams EF and GH, and the horizontal beam IM, which is simply supported at I and M, rests on the other beams at J and L, and carries the vertical load 2 P at K. Each beam is to have uniform cross section, and the beams EF and GH are to be identical. Subregion V_1 thus is the segment IM, and subregion V_2 consists of

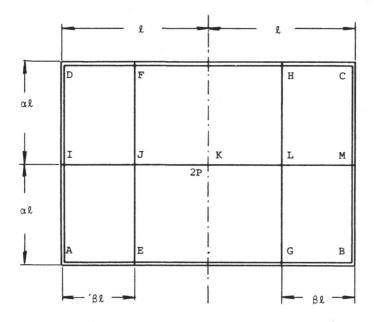

Fig. 4.1: Simply supported beams spanning a rectangular opening; beam IM also rests on beams EF and GH. Load 2P acts at K.

the segments EF and GH. As design parameters, we use the yield moments Y_1
and Y_2 in these subregions. We do not set bounds on these yield moments,
but note that, even in the absence of explicit bounds, Y_2 is bounded from
below by zero. The specific costs will be assumed to be proportional to
the yield moments, so that the total cost is proportional to

$$\Omega = 2\,Y_1 + 4\,\alpha\,Y_2 \,. \tag{4.26}$$

Figure 4.2a shows a possible collapse mechanism for subregion V_1 with
yield hinges at J, K, and L. The corresponding mechanism for

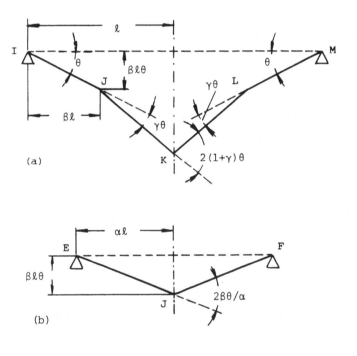

(a)

(b)

Fig. 4.2: Collapse mechanisms. The ordinates indicate rates
of deflection, and the slopes, rates of rotation.

subregion V_2 is shown in Fig. 4.2b; it has yield hinges at J and L. The
internal power of dissipation for these mechanisms is

$$D = 2 (1 + 2\gamma) \theta Y_1 + 4 \beta \theta Y_2 / \alpha . \tag{4.27}$$

Because $Y_1 > 0$, optimality requires that

$$(1 + 2\gamma) \theta = 1 , \tag{4.28}$$

$$\beta \theta / \alpha \begin{cases} \leq \alpha & \text{for } Y_2 = 0 , \\ = \alpha & \text{for } Y_2 > 0 . \end{cases} \tag{4.29}$$

We discuss first the case $Y_2 > 0$. Elimination of θ between (4.28) and the second relation (4.29), and solution of the resulting equation for γ furnishes

$$\gamma = (\beta - \alpha^2) / (2\alpha^2) . \tag{4.30}$$

Figure 4.3 shows the forces acting on the beams; the magnitude of the force the beams IM and EF (or GH) exert on each other at J (or L) has been written as ρP . The bending moments of beam IM at J and K are

$$Q_J = (1 - \rho) \beta P \ell , \quad Q_K = (1 - \rho \beta) P \ell . \tag{4.31}$$

For the yield mechanism shown in Fig. 4.2a to develop, we must have $-Q_J = Q_K = Y_1$ or

$$\rho = (1 + \beta) / (2\beta) . \tag{4.32}$$

With this value of ρ , we find

$$Y_1 = (1 - \rho \beta) P \ell = (1 - \beta) P \ell / 2 ,$$

$$Y_2 = \alpha \rho P \ell / 2 = \alpha (1 + \beta) P \ell / (4\beta) . \tag{4.33}$$

For this analysis to be valid, γ as given by (4.30) must be nonnegative, or

$$\beta \geq \alpha^2 . \tag{4.34}$$

In the limiting case $\beta = \alpha^2$, our analysis furnishes

$$\rho = (1 + \alpha^2) / (2 \alpha^2) ,$$

$$(4.35)$$

$$Y_1 = (1 - \alpha^2) P \ell / 2 , \quad Y_2 = \alpha (1 + \alpha^2) P \ell / (4 \alpha^2)$$

and hence

$$\Omega = 2 P \ell$$

by (4.26).

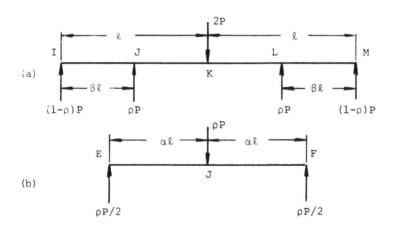

Fig. 4.3: Forces acting on beams; ρP is force exerted
by beam IM on beams EF and GH

In the case $Y_2 = 0$, we have $\rho = 0$, and hence $\gamma = 0$. Elimination of θ
between (4.28) and the first relation (4.29) then furnishes

$$\beta \leq \alpha^2$$

$$(4.37)$$

as the characteristic of this case. With $\rho = 0$, we have

$$Y_1 = P \ell , \qquad Y_2 = 0 , \tag{4.38}$$

and

$$\Omega = 2 P \ell$$

by (4.36). In the limiting case $\beta = \alpha^2$, the designs (4.35) and (4.38) thus have the same weight, and so has any design corresponding to a value of ρ between $\rho = 0$ and the value in (4.35). For $\beta < \alpha^2$ and $\beta > \alpha^2$, however, the optimal design is unique.

4.3. Optimal division into elements of prescribed shape. To illustrate the treatment of this kind of problem, let us consider a horizontal sandwich beam with uniform rectangular core section that is built in at $x = 0$ and simply supported at $x = 2 \ell$, and carries a vertical load 2 P at $x = \ell$ (Fig. 4.4a). We set $\xi = x / \ell$ and divide the span into the segments $0 \le \xi < \beta < 1$ and $\beta < \xi \le 2$, where β will at first be regarded as given. In each of these segments, the yield moment is to have a constant value, and these values, Y_1 and Y_2 , will be used as the design parameters.

Figures 4.4b and 4.4c show possible collapse mechanisms that have yield hinges at $\xi = 0$, $\xi = 1$, and $\xi = \beta + 0$ (Fig. 4.4b) or $\xi = \beta - 0$ (Fig. 4.4c). The internal power of dissipation is

$$D = \begin{cases} \theta \, Y_1 + (2 + 3 \, \gamma - \beta \, \gamma) \, \theta \, Y_2 \text{ for Fig. 4.4b,} \\ (1 + \gamma) \, \theta \, Y_1 + (2 - 2 \, \gamma + \beta \, \gamma) \, \theta \, Y_2 \text{ for Fig. 4.4c.} \end{cases} \tag{4.40}$$

The weight of the cover plates, which is to be minimized, is proportional to

$$\Omega = \beta \, Y_1 \, \ell + (2 - \beta) \, Y_2 \, \ell . \tag{4.41}$$

Since optimality requires that $\partial D / \partial Y_i = \partial \Omega / \partial Y_i$ for i = 1, 2, and since γ as shown in Figs. 4.4b and 4.4c must be positive, we find

$$\theta = \beta\ell \ , \quad \gamma\theta = (2-3\beta)\ell/(3-\beta) \quad \text{for } \beta < 2/3 \ ,$$

$$(4.42)$$

$$\theta = (2+\beta-\beta^2)\ell/(4-\beta) \ , \quad \gamma\theta = (3\beta-2)\ell/(4-\beta) \quad \text{for } \beta > 2/3 \ .$$

If the yield mechanisms in Figs. 4.4b and 4.4c are to occur, the bending

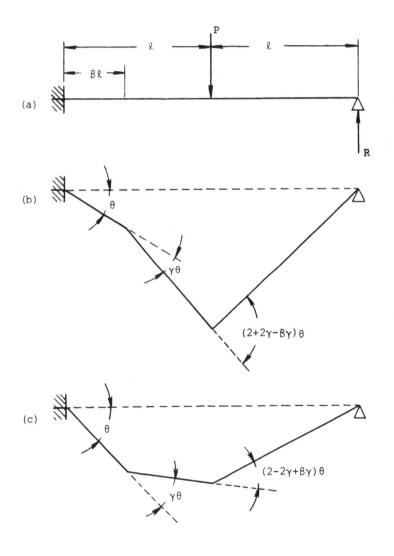

Fig. 4.4: (a) beam; (b) and (c) collapse mechanisms

moment $Q(\xi)$ must satisfy the relations

$$Q(0) = -Y_1 , \qquad Q(\beta) = -Q(1) = -Y_2 \quad \text{for } \beta < 2/3 ,$$

$$\text{(4.43)}$$

$$Q(0) = -Q(\beta) = -Y_1 , \qquad Q(1) = Y_2 \quad \text{for } \beta > 2/3 .$$

Expressing $Q(\xi)$ in terms of P and the reaction R at $\xi = 2$, and using (4.43), we finally obtain the optimal designs

$$Y_1 = (1 + \beta) P \ell / (3 - \beta)$$
$$\left. \vphantom{\begin{array}{c}a\\b\end{array}} \right\} \quad \text{for } \beta < 2/3 , \qquad \text{(4.44a)}$$
$$Y_2 = R \ell = (1 - \beta) P \ell / (3 - \beta)$$

$$Y_1 = \beta P \ell / (4 - \beta)$$
$$\left. \vphantom{\begin{array}{c}a\\b\end{array}} \right\} \quad \text{for } \beta > 2/3 . \qquad \text{(4.44b)}$$
$$Y_2 = R \ell = (2 - \beta) P \ell / (4 - \beta)$$

For these designs, the objective function (4.41) is found to be

$$\Omega(\beta) = \begin{cases} 2 (1 - \beta + \beta^2) P \ell^2 / (3 - \beta) & \text{for } \beta < 2/3 , \\ 2 (2 - 2\beta + \beta^2) P \ell^2 / (4 - \beta) & \text{for } \beta > 2/3 . \end{cases} \qquad \text{(4.45)}$$

As β tends towards 2/3 from below or above, the reactions R furnished by the second equations (4.44a) and (4.44b) are $R = P/7$ and $R = 2P/5$. For $\beta = 2/3$, the designs

$$Y_1 = (P - 2 R) \ell , \qquad Y_2 = R \ell \qquad \text{(4.46)}$$

furnish $\Omega = 2P\ell/3$ for any value of R satisfying $1/7 \leq R/P \leq 2/5$.

So far, we have regarded β as given in advance. The formulas (4.45), however, are readily used to determine the value of β that minimizes $\Omega(\beta)$. One finds that $\Omega(\beta)$ has a minimum in each of the intervals $0 < \beta < 2/3$ and $2/3 < \beta < 1$. The first minimum, which corresponds to $\beta = 3-\sqrt{7} = 0.35425$, has the smaller value, namely $\Omega = 0.58301\ P\ell^2$. Note that this is about 12.5% smaller than the maximal value $\Omega = 2P\ell^2/3$, which is assumed for $\beta = 0, 2/3,$ and 1.

The use of a condition derived by Rozvany[35] somewhat shortens the deter-
mination of the β-value that minimizes $\Omega(\beta)$. Imagine the discontinuous
change of the yield moment at $\xi = \xi_0$ replaced by a gradual transition
from Y_1 at $\xi = \xi_0 - \varepsilon$ to Y_2 at $\xi = \xi_0 + \varepsilon$. As the length 2ε of this
transitional segment tends to zero, the contributions of the segment to D
and Ω as defined in (4.40) and (4.41) tend to $(Y_1 + Y_2) \varepsilon |q|$ and
$(Y_1 + Y_2) \varepsilon$, where q is the mean rate of curvature of the segment. Op-
timality requires $|q| = 1$ or $q = sgn\ Q(\xi_0)$. The increase Δr in the rate
of rotation from $\xi = \xi_0 - \varepsilon$ to $\xi = \xi_0 + \varepsilon$ thus equals

$$\Delta r = 2\ \varepsilon\ sgn\ Q(\xi_0) , \tag{4.47}$$

while the increase in bending moment may be written as

$$\Delta Q = \Delta Y\ sgn\ Q(\xi_0) = 2\ \varepsilon\ S , \tag{4.48}$$

where S is the shear force at $\xi = \xi_0$ and $\Delta Y = Y_2 - Y_1$. Elimination of
ε between (4.47) and (4.48) furnishes

$$\Delta r = \Delta Y\ /\ S . \tag{4.49}$$

To illustrate the use of this condition, we reconsider the problem in
Fig. 4.4a. From equations (4.44a), we find

$$\Delta Y = -2\ \beta\ P\ \ell\ /\ (3 - \beta) ,$$
$$\tag{4.50}$$
$$S = P - R = 2\ P\ /\ (3 - \beta) .$$

The condition (4.49) thus yields

$$\Delta r = -\ \beta\ \ell . \tag{4.51}$$

Equating this value of Δr to the value $-\gamma\ \theta$ in the first line of (4.42),
we again obtain $\beta = 3 - \sqrt{7}$ as the optimal value of β .

5. Optimal Layout of Trusses

5.1. Optimal layout for a single state of loading. In Section 3.2b we discussed optimal plastic design of a truss of given layout. The notations and results of that section will now be used in the discussion of the following problem. A plane truss is to transmit the given load P to a given rigid foundation, which is indicated by shading in Fig. 5.1. The bars of the truss are to be made from a rigid, perfectly plastic material with tensile and compressive yield limits $\pm\sigma_0$. The given load is to represent the load-carrying capacity of the truss, and the total volume of the bars of the truss is to be minimized. Note that the choice of the layout of the truss is left to the designer, except that there must be a joint at the given point of application of the load, and that joints on the foundation arc, wich represents the surface of the rigid foundation, are restrained from moving.

Once the potential joints have been chosen, the layout of the basic truss is completely determined, and the results of Section 3.2b become applicable. To enable us to omit the typical bar i of the truss, we set the lower bound $\bar{V}_i$ on the volume of the bar equal to zero. According to (3.28), the optimal truss then admits a collapse mechanism satisfying

$$|q_i| \begin{cases} = q_0 & \text{if bar i is retained,} \\ \\ \leq q_0 & \text{if bar i is omitted.} \end{cases} \tag{5.1}$$

Here, q_0 is an arbitrary reference strain rate for all bars of the basic truss, and q_i is the axial strain rate in bar i of this truss as determined from the velocities of its endpoints in the considered collapse mechanism.

In the limit of a uniformly dense distribution of potential joints, the optimality condition (5.1) stipulates a velocity field of collapse such

that, at any potential joint J, the strain rates in the directions of the
bars through J have the absolute values q_0 , while the strain rate for no
other direction at J exceeds q_0 in absolute value. If some bars at J are
stressed to the tensile, and others to the compressive yield limit, the
principal strain rates of the collapse field at J must have the values
$\pm q_0$, and the bars through J must be in the principal directions, which
form right angles. It will now be shown that this kind of velocity field
of collapse is uniquely determined by the foundation arc and by the fact
that the principal strain rates have the values $\pm q_0$ throughout the field.

Denoting differentiation with respect to the rectangular Cartesian coor-
dinates x_1 , x_2 in the plane of the truss by ∂_1 and ∂_2 , let p_1 and p_2 be
the velocity components of the field of collapse,

$$q_1 = \partial_1 p_1 , \quad q_2 = \partial_2 p_2 , \quad q_3 = (\partial_1 p_2 + \partial_2 p_1)/2 \qquad (5.2)$$

the rates of extension and shear in the coordinate directions, and

$$r = (\partial_1 p_2 - \partial_2 p_1)/2 \qquad (5.3)$$

the mean rate of rotation. If the angle between the negative x_2-direction
and the arbitrarily assigned positive direction along the line with the
strain rate q_0 is denoted by θ , then

$$q_1 = -q_0 \cos 2\theta , \quad q_2 = q_0 \cos 2\theta , \quad q_3 = -q_0 \sin 2\theta . \qquad (5.4)$$

With $r' = r/2q_0$, it follows from (5.2) through (5.4) that

$$\partial_1 p_1 = -q_0 \cos 2\theta , \quad \partial_2 p_1 = -q_0 (2r' + \sin 2\theta) ,$$
$$(5.5)$$
$$\partial_1 p_2 = q_0 (2r' - \sin 2\theta) , \quad \partial_2 p_2 = q_0 \cos 2\theta .$$

Elimination of p_1 and p_2 from (5.5) by cross-differentiation furnishes

$$\partial_1 r' + \partial_1 \theta \cos 2\theta + \partial_2 \theta \sin 2\theta = 0 ,$$
$$(5.6)$$
$$\partial_2 r' + \partial_1 \theta \sin 2\theta - \partial_2 \theta \cos 2\theta = 0 .$$

Let the axes of x_1 and x_2 be chosen parallel to the directions of the principal strains q_0 and $-q_0$ at the typical point J of the velocity

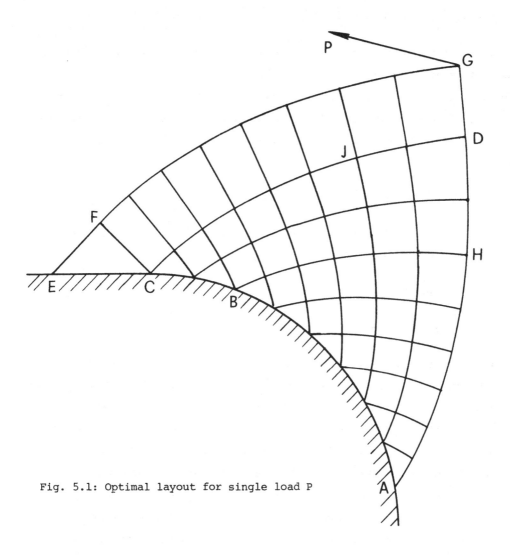

Fig. 5.1: Optimal layout for single load P

field of collapse, and denote differentiation in these directions by d_1 and d_2 . At the considered point, $\theta = \pi/2$ and equations (5.6) take the form

$$d_1 (r' - \theta) = 0 , \quad d_2 (r' + \theta) = 0 . \tag{5.7}$$

Since each of these equations only involves differentiation in a single direction, the lines of principal strain are the characteristics of the hyperbolic system (5.6).

At the typical point B of the foundation arc, let the positive direction along this arc form the angle ϕ with the positive x_1-direction. Since the foundation arc is rigid, the strain rate along the arc vanishes. Accordingly, the tangent and normal of the foundation arc at B bisect the right angles formed by the directions of the principal strain rates at B. Thus,

$$\theta = \phi + \pi/4 \quad \text{or} \quad \theta = \phi + 3\pi/4 . \tag{5.8}$$

Since p_1 vanishes along the foundation arc, $(\partial_1 p_1) \cos \phi + (\partial_2 p_1) \sin \phi = 0$ at B. With the use of (5.5) and (5.8), this condition yields

$$r' = \pm \, 1/2 \tag{5.9}$$

along the foundation arc.

From the initial conditions (5.8) and (5.9), which are of Cauchy type, and the characteristic equations (5.7), the net of characteristics can be constructed in the well-known manner (see, for instance, Prager and Hodge[36], pp. 134 - 139). In Fig. 5.1, only a few members of each family of characteristics are shown. The characteristics GA and GE through the point of application G of the load P form the contour of the optimal truss. Of course, the considered type of collapse field with principal strain rates of opposite signs is relevant to the solution of the problem of optimal layout only if the direction of the load P is such that the statically determinate forces in the bars through G have opposite signs.

In regions such as ACD, where the characteristics of both families are

curved, the optimal truss consists of two dense, orthogonal families of
bars of infinitesimal length, and thus is a trusslike continuum rather
than a truss in the usual sense of the term. In regions such as CDGF,
where the characteristics of one family are straight, there are no interior members orthogonal to the straight characteristics. In regions such
as CFE, where the characteristics of both families are straight, there
are no interior members.

The construction of a velocity field of collapse that satisfies the optimality condition (5.1) in the region AGE establishes the fact that no
other arrangement of joints and bars <u>in this region</u> can furnish a truss
of lesser weight, but leaves the possibility that locating some joints
outside this region might result in a lighter truss. This possibility,
however, is excluded if the considered collapse field can be continued
beyond the contour AGE of the optimal truss in such a manner that the optimality condition (5.1) remains satisfied throughout the region that is
available for additional or alternative joints. In Fig. 5.1, this kind of
continuation is clearly possible.

Trusslike continua of the kind considered here were first discussed by
Michell[37], who investigated optimal design of trusses for given allowable
stresses.

It is worth noting that the equations (5.6) have the same form as the basic equations of the slip line field in plane flow of a rigid, perfectly
plastic solid (see, for instance, Prager and Hodge[36], p.130). Accordingly, the bars of the optimal truss form a Hencky-Prandtl net, and the numerical and graphical methods that have been developed for the construction
of this kind of net are applicable to the present problem (see, for instance, Hill[38], p.140 ff., and Prager[39]). Of the many remarkable geometrical properties of Hencky-Prandtl nets, we mention only one. The tan-
gents of two arbitrary Hencky-Prandtl lines of one family at their points
of intersection with a line of the other family form an angle that is in-

dependent of the choice of the latter line. We shall refer to this as the basic geometric property of Hencky-Prandtl nets.

So far, it has been assumed that the direction of the load P is such that on of the bars at the point of application of the laod is stressed in tension, and the other, in compression. Figure 5.2 shows an interesting

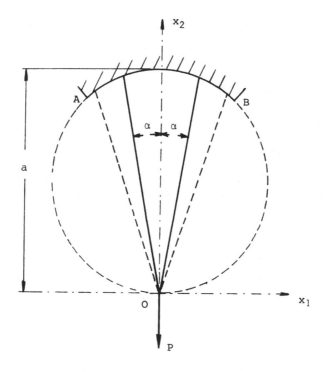

Fig. 5.2: Nonunique optimal layout

case, in which this condition is not fulfilled. The "foundation arc" AB is part of a circle of diameter a, which passes through the point O at which the vertical load P is acting. With respect to the rectangular co-ordinates x_1, x_2 shown in Fig. 5.2, let

$$p_1 = 0 , \qquad p_2 = v_0 \{ (x_1^2 + x_2^2) / (ax_2) - 1 \} \qquad\qquad (5.10)$$

be the velocity components of a proposed field of collapse. The corresponding strain rates are

$$q_1 = \partial_1 p_1 = 0 , \qquad q_2 = \partial_2 p_2 = v_0 (x_2^2 - x_1^2) / (ax_2) ,$$

$$\qquad\qquad\qquad\qquad\qquad\qquad\qquad\qquad\qquad\qquad\qquad (5.11)$$

$$q_3 = (\partial_1 p_2 + \partial_2 p_1) / 2 = v_0 x_1 / (ax_2) ,$$

and the principal strain rates are found to be

$$q_{max} = v_0 / a , \qquad q_{min} = -v_0 x_1^2 / (ax_2^2) \qquad\qquad (5.12)$$

in the direction of the radius vector from O and normal to it. Within the
region OAB in Fig. 5.2, q_{min} thus has a smaller absolute value than q_{max}.
The optimality condition (5.1) is therefore fulfilled with $q_0 = v_0 / a$,
provided that all bars run radially from O to the foundation arc.

Consider, for instance, a truss consisting of two bars from O to the
foundation arc that form angles $\pm \alpha$ with the x_2-axis (full lines in Fig.
5.2). Tensile forces in these bars that equilibrate the load P have the
magnitude P/(2cosα) and therefore need the cross-sectional area A =
P/(2 σ_0 cosα). On the other hand, the length of each of these bars is
ℓ = a cosα . The total volume of the two bars, 2Aℓ = P/(σ_0a) , thus is
independent of α. This means that one may regard the load P as the sum of
two vertical, downward loads P' and P", attribute P' to the full-line
bars in Fig. 5.2 and P" to the broken-line bars, and determine the cross-
sectional areas of each pair of bars so as to have all bars at the tensile yield stress σ_0. The truss consisting of the four bars has again the
total bar volume P/(σ_0a) regardless of the division of P into P' and P".

This lack of uniqueness of the optimal layout is caused by the fact that
the point of application O of the load lies on the circle of the foundation arc. When O is inside this circle, the optimal "truss" consists of

a single bar along the line of action of the load. If, on the other hand,
O is outside the circle of the foundation arc, the optimal truss consists
of the two bars joining O to the endpoints A and B of the foundation arc.

In concluding this section, we remark that the optimal design of an elastic truss for prescribed compliance to the given load is characterized by
an optimality condition that is obtained from (5.1) by replacing $|q_i|$ by
q_i^2 . It follows that the optimal layout discussed in connection with plastic design is also optimal in elastic design for given compliance, as
was remarked by Hegemier and Prager[40], who also discussed other problems
for which this layout is optimal.

5.2. Optimal layout for two alternative loadings. An interesting superposition principle that furnishes the optimal truss for two alternative
loadings was established by Hemp[41]. The illustrative problem in Fig. 5.3
concerns the transmission of the alternative loads P' and P" to the sha-
ded rigid foundation arc by means of a truss of minimal weight that is
to be at the limit of its load-carrying capacity under either one of the
loads. The optimality condition (5.1) must now be replaced by

$$|q_i'| + |q_i''| \begin{cases} = q_0 \text{ if bar } i \text{ is retained,} \\[2mm] \leq q_0 \text{ if bar } i \text{ is omitted .} \end{cases} \tag{5.13}$$

Here, q_0 is again an arbitrary reference strain rate for all bars of the
basic truss, while q_i' and q_i'' are the axial strain rates of bar i of this
truss in collapse mechanisms for the two loads.

It will now be shown that the optimal truss for the alternative loads P'
and P" is obtained by superposition of the optimal trusses for the single
loads $\bar{P} = (P'+P'')/2$ and $\bar{\bar{P}} = (P'-P'')/2$ (Fig. 5.3). The terms component
loads, component trusses, and component fields will be used to indicate
the loads $\bar{P}$ and $\bar{\bar{P}}$, the optimal trusses for them, and the velocity fields
of their collapse mechanisms.

A line element of the plane of the optimal truss will be said to be of
the first, second, or third kind, depending on whether it is along a bar
of the first component truss, a bar of the second component truss, or

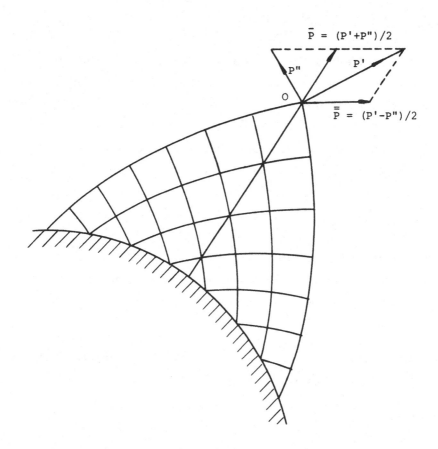

Fig. 5.3: Optimal truss for alternative loads P' and P" is obtained
by superposition of optimal trusses for single loads (P'±P")/2

not along a bar of either component truss. If $\bar{q}$ and $\bar{\bar{q}}$ are the strain rates of the same line element in the two component fields, the optimality
condition (5.1) requires that

$$\left.\begin{array}{l} |\bar{q}| \leq q_0 \\[2mm] |\bar{\bar{q}}| \leq q_0 \end{array}\right\} \quad \begin{array}{l} \text{with equality for} \\ \text{line elements of the} \end{array} \quad \left\{\begin{array}{l} \text{first} \\[2mm] \text{second} \end{array}\right\} \text{ kind.} \qquad (5.14)$$

It follows from (5.14) that

$$|\bar{q}+\bar{\bar{q}}| + |\bar{q}-\bar{\bar{q}}| \leq 2q_0 \left\{\begin{array}{l} \text{with equality for line elements} \\ \text{of the first or second kinds.} \end{array}\right. \qquad (5.15)$$

The sum and the difference of the component fields thus satisfy the optimality condition for the truss obtained by the superposition of the component trusses (with reference strain rate $2q_0$), while the sum Q_i' and
the difference Q_i'' of the bar forces $\bar{Q}_i$ and $\bar{\bar{Q}}_i$ of the component trusses
are in equilibrium with the given alternative loads $P' = \bar{P} + \bar{\bar{P}}$ and $P'' =$
$\bar{P} - \bar{\bar{P}}$. These remarks establish the superposition principle provided
that, for each bar i of the truss obtained by the superposition, the for-
ces $Q_i' = \bar{Q}_i + \bar{\bar{Q}}_i$ and $Q_i'' = \bar{Q}_i - \bar{\bar{Q}}_i$ do not have signs opposite to those of
the strain rates $q_i' = \bar{q}_i + \bar{\bar{q}}_i$ and $q_i'' = \bar{q}_i - \bar{\bar{q}}_i$. It will now be shown
that this condition is fulfilled. In the following discussion, it is im-
portant to note that, when the axial strain rate of a bar vanishes, the
bar force my have any value from the compressive to the tensile yield
force.

The component trusses are statically determinate. Accordingly, the force
in bar i of the first component truss is given by

$$\bar{Q}_i = \sigma_0 \bar{A}_i \text{ sgn } \bar{q}_i , \qquad (5.16)$$

where $\bar{A}_i$ is the cross-sectional area of the bar. Similiarly, we have

$$\bar{\bar{Q}}_j = \sigma_0 \bar{\bar{A}}_j \text{ sgn } \bar{\bar{q}}_j \qquad (5.17)$$

for bar j of the second component truss. Several cases must be distingui-
shed as follows.

    (a) If $\bar{A}_i > 0$, $\bar{\bar{A}}_i = 0$, we have $\bar{\bar{Q}}_i = 0$ and $Q_i' = Q_i'' = \bar{Q}_i$. Moreover,
        $|\bar{q}_i| = q_0$, $|\bar{\bar{q}}_i| \leq q_0$. Accordingly, q_i' and q_i'' will have the

sign of $\bar{q}_i$ or vanish. Thus, Q'_i and Q''_i cannot have signs opposite to those of q'_i and q''_i .

(b) If $\bar{A}_i = 0$, $\bar{\bar{A}}_i > 0$, we have $\bar{Q}_i = 0$ and $Q'_i = -Q''_i = \bar{\bar{Q}}_i$. Moreover,
$|\bar{q}_i| \le 0$, $|\bar{\bar{q}}_i| = q_0$. Accordingly, q'_i and q''_i will have the
signs of $\bar{\bar{q}}_i$ or $-\bar{\bar{q}}_i$ or vanish. Thus, Q'_i and Q''_i cannot have signs
opposite to those of q'_i and q''_i .

(c) If $\bar{A}_i > 0$, $\bar{\bar{A}}_i > 0$, we have $|\bar{q}_i| = |\bar{\bar{q}}_i| = q_0$. If, for instan-
ce, $\bar{q}_i = -\bar{\bar{q}}_i = q_0$, we have $q'_i = 0$, and the sign of Q'_i does not
matter. Moreover, $\bar{Q}_i = \sigma_0 \bar{A}_i$, $\bar{\bar{Q}}_i = -\sigma_0 \bar{\bar{A}}_i$, and $Q''_i = \sigma_0 (\bar{A}_i + \bar{\bar{A}}_i)$
has the same sign as $q''_i = 2q_0$. The other sign combinations for
the strain rates $\bar{q}_i$ and $\bar{\bar{q}}_i$, which have the absolute value q_0 ,
can be discussed in similar manner. In all cases, it is seen
that Q'_i and Q''_i cannot have signs opposite to those of q'_i and q''_i.

The foregoing proof of the superposition principle is due to Nagtegaal
and Prager[42]. The original proof of Hemp[41] was based on a linear program-
ming formulation of the problem.

5.3. Discretization of Michell continua. Michell's trusslike continua
achieve an absolute minimum of the total weight of bars, but are not
practical structures because they have infinitely many bars and joints.
A way of enforcing finite numbers of bars and joints is to include the
weight of connections (rivets and gusset plates) in the structural
weight that is to be minimized. One may, for instance, assume that the
weight of the connections required by bar i is proportional to the force
in this bar, that is, proportional to its cross-sectional area A_i . The
objective function then is

$$\Omega = \sum_i A_i (\ell_i + \ell_0) , \tag{5.18}$$

where ℓ_0 does not depend on i. and the optimality condition is found to be

$$\ell_i |q_i| \le (\ell_i + \ell_0) q_0 \quad \text{with equality for } A_i > 0 \tag{5.19}$$

(Prager[43]). Here, q_0 is again a reference strain rate, and q_i is the axial strain rate of bar i in a collapse mechanism of the optimal truss. Note that, for $\ell_0 = 0$, the condition (5.19) reduces to (5.1).

As example for the use of the optimality condition (5.19), consider the transmission of a horizontal load P from its point of application O to a straight, horizontal foundation line of unrestricted length that has the distance h from O. Since the optimal truss will be symmetric with respect to the vertical through O, this joint will have a horizontal velocity p in the collapse mechanism of the optimal truss under the load P. Because the time scale of the collapse does not matter, p may be assigned a numerical value equal to that of h. If the bar i forms the angle θ_i with the vertical, its length is $\ell_i = h \, / \cos \theta_i$, and its strain rate q_i in the considered collapse mechanism has the absolute value

$$|q_i| = (h/\ell_i) \sin \theta_i = 0.5 \sin 2\theta_i . \qquad (5.20)$$

The optimality condition (5.19) then furnishes

$$0.5 \sin 2\theta_i - (\ell_0 q_0/h) \cos \theta_i \leq q_0 . \qquad (5.21)$$

The bars of the optimal truss thus form angles $\theta_i = \pm\theta$ with the vertical that maximize the left side of (5.21). Accordingly,

$$\ell_0 \, q_0 \, / \, h = - \cos 2\theta \, / \sin \theta . \qquad (5.22)$$

Elimination of q_0 between (5.21) and (5.22) yields the equation

$$\ell_0 \, / \, h = -\cos 2\theta \, / \cos^3 \theta \qquad (5.23)$$

from which θ may be found by iteration when ℓ_0/h is given. For example, for $\ell_0/h = 0.4$, one finds $\theta = 48.37^{\circ}$. The corresponding value of the objective function (5.18) is $\Omega = 2.550 \; Ph/\sigma_0$. For a two-bar truss whose bars form the angles $\pm 45^{\circ}$ with the vertical, the objective function (5.18) is found to have the value $\Omega^* = 2.566 \; Ph/\sigma_0$. Now, with $\ell_0/h = 0.4$, the bars of this truss contribute an amount equal to about 28 % of their weight to the weight of connections. Even for these rather heavy connections, Ω^* exceeds the optimum Ω by only 0.63 %. This example suggests a rather low response of the structural weight to deviations from the optimal layout.

The next example differs from the preceding one only by the fact that the foundation has a finite width 2 b. If the optimal two-bar truss considered above is ruled out by the given value of h, the optimal layout is of the kind shown in Fig. 5.4. To prove that a specific layout L of this type is optimal for a given value of ℓ_0/h , we might proceed as follows. Having chosen a value of q_0 , we assign to the typical bar i the rate of elongation $\lambda_i = (\ell_i + \ell_0) q_0$, and determine the corresponding joint velocities, for instance by a Williot diagram. In each triangular or quadrilateral mesh formed by bars, we then determine a linear or bilinear velocity field based on the velocities of the vertices of the mesh. Finally, we continue this field beyond the contour AOB of the truss.

Next, we consider an alternative layout L* obtained from L by displacing its joints, while maintaining symmetry and avoiding any crossing of bars. For each bar of L*, we determine the rate of elongation λ_i^* as the integral over the rates of elongation of its elements in the considered collapse field. The layout L is optimal if $\lambda_i^* \le (\ell_i + \ell_0) q_0$ for each bar of the layout L* and for all possible positions of its joints. To verify this would obviously be a most difficult task. Instead of searching for the truly optimal layout, we may therefore consider nearly optimal layouts obtained by appropriate discretization of the Michell layout.

Consider the bars in the left contour OGFB of the truss in Fig. 5.4. The forces in the corresponding bars of the Michell layout have a constant magnitude, while the forces in the bars normal to the contour vanish. We shall stipulate the first property for the contour bars of the layout in

Fig. 5.4. Since the second property cannot be stipulated for this layout, we shall instead demand that the forces in the bars that go from contour joints into the interior of the layout are to have the same intensity. It follows that (1) the angle α between the contour bars at a contour joint

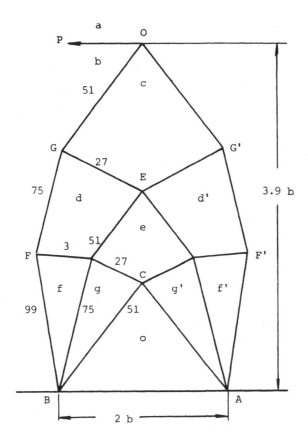

Fig. 5.4: Discretized Michell structure

is bisected by the third bar through this joint, and (2) the angle α has the same value at each contour joint. Let the properties (1) and (2) also be stipulated for the bars of the chain BDEG' and the bars DC and ED', as well as the bars of the chain BCD'F' and the bars CA and D'A, and the chains and bars obtained from these by symmetry with respect to the vertical through O. In each quadrilateral mesh formed by bars, opposite sides then form the angle $\pi - \alpha$, and this may be regarded as a suitable discretization of the basic property of Hencky-Prandtl nets.

In Fig. 5.4, the numbers appearing next to the bars give their angles with the horizontal (in degrees); they will enable the reader to verify that the layout has the geometric properties stipulated above.

Palmer and Sheppard[44] have defined the _Michell efficiency_ of a truss as the inverse ratio between its total volume of bars and that of the Michell structure for the same loads. For the truss in Fig. 5.4, this efficiency is found to be 97.8 % (Prager[43]). This high value suggests that the proposed discretization of Michell structures does not entail a significant loss in efficiency.

6. Optimal Layout of Grillages

6.1. Basic types of collapse fields. This chapter is devoted to the optimal design of a grillage of horizontal beams that are to transmit a given vertical, downward loading to given supports. The beams are to have
rectangular cross sections of a common uniform height but variable width.
The given loading is to be at the load-carrying capacity of the grillage,
and the total volume of the beams is to be minimized.

Figure 6.1 shows a typical problem: a square opening ABCD is to be cover-
ed by a grillage of beams that is simply supported along the edges of the
square. An optimal layout of beams, which is not unique, is shown in Fig.
6.1a (Morley[45], Rozvany[46]). All beams are parallel to AC or BD. The
beams in the central square EFGH are simply supported by the beams that
form the edges of this square. All beams in a corner triangle, such as
AEH, have the direction of the diagonal of the square ABCD that does not
pass through this corner. These beams are simply supported at the edges
of the square ABCD. It will be shown later that the total volume of the
beams of the layout is 37.5 % less than the volume of the beams of a
grillage formed by beams parallel to the sides of the square ABCD.

Similarly to the problem of the optimal layout of a truss, the problem of
the optimal layout of a grillage may be approached by starting from a
pattern of potential junctions of beams, forming the basic grillage, in
which any two junctions are connected by a beam, and discussing which
beams should be omitted in the optimal layout. In the limit of a uniformly dense distribution of junctions, this approach leads to an optimality condition that is analogous to the one obtained in Section 5.1. The
optimal grillage admits a collapse mechanism with a rate of deflection
field that satisfies the kinematic conditions at the supports and has

principal rates of curvature that do not exceed a given reference rate
of curvature q_0 in absolute value. Along any beam of the optimal grillage, the rate of curvature of the collapse field must have the absolute
value q_0 , and the bending moment cannot have a sign opposite to that of
the rate of curvature.

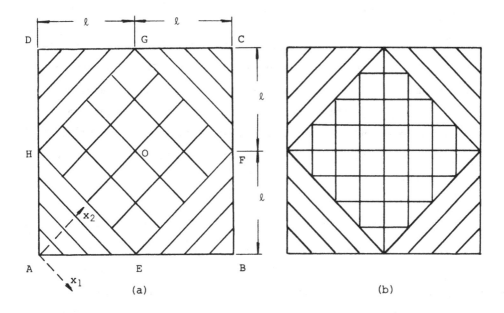

Fig. 6.1: Optimal layouts for simply supported square grillage

We shall denote the principal rates of curvature of the collapse field
of an optimal grillage by q_1 and q_2 , and the corresponding bending moments by Q_1 and Q_2 , and label the principal directions in such a manner
that $|q_1| = q_0$. Depending on whether $|q_2| < q_0$ or $|q_2| = q_0$, we then
distinguish the following basic types of regions in the collapse field:

Type

R^+: $q_1 = q_0$

R^-: $q_1 = -q_0$ $\left.\right\}$ $|q_2| < q_0$, $Q_1 \text{ sgn } q_1 \geq 0$, $Q_2 = 0$,

S^+: $q_1 = q_2 = q_0$

S^-: $q_1 = q_2 = -q_0$ $\left.\right\}$ $Q_1 \text{ sgn } q_1 \geq 0$, $Q_2 \text{ sgn } q_2 \geq 0$,   (6.1)

T: $q_1 = -q_2 = q_0$, $Q_1 \geq 0$, $Q_2 \leq 0$.

It turns out that, for the majority of straightedged grillages, the collapse for the optimal layout has a rate of deflection field that is regionwise quadratic in the rectangular coordinates in the plane of the grillage. In a region of this kind, the principal rates of curvature have fixed directions, and we shall take the axes of x_1 and x_2 in these directions. In a region of type T, for instance, the collapse field then has a rate of deflection of the form

$$p = -(q_0/2) \; (\; x_1^2 - x_2^2 \;) + a \; x_1 + b \; x_2 + c \; , \qquad (6.2)$$

where a, b, and c are constants. Note that, for a = b = c = 0, we have $p = 0$ for $x_1 \pm x_2 = 0$. The field (6.2) thus is appropriate in the neighborhood of the corner of a rectangular plate that is simply supported along the bisectors of the first and second quadrants. In the grillages of Fig. 6.1, the corner triangles, such as AEH, have collapse fields of this type. According to (6.2), the principal curvatures have the values

$$q_1 = -\partial_{11}p = q_0 \; , \quad q_2 = -\partial_{22}p = -q_0 \; . \qquad (6.3)$$

This means that we have the rate of curvature q_0 along EH, and hence also along EF, FG, and GH. Inside the square EFGH, we may therefore expect a collapse field of type S^+, that is rates of deflection of the form

$$\bar{p} = -(q_0/2) \; (\; x_1^2 + x_2^2 \;) + \bar{a} \; x_1 + \bar{b} \; x_2 + \bar{c} \; . \qquad (6.4)$$

For reasons of symmetry, we must have $(\; \partial_1 + \partial_2 \;) \; \bar{p} = 0$ along EG, that is

$x_1 + x_2 = \ell \sqrt{2}$, and $(\partial_1 - \partial_2) \bar{p} = 0$ along FH, that is for $-x_1 + x_2 = \ell \sqrt{2}$. Moreover, at E, that is for $x_1 = x_2 = \ell/\sqrt{2}$, we must have $\bar{p} = 0$. In view of these conditions, (6.4) takes the form

$$\bar{p} = -(q_0/2) (x_1^2 + x_2^2 - 2\sqrt{2} \ell x_2 + \ell^2) . \qquad (6.5)$$

Our guess that the collapse field is of type S^+ throughout the square EFGH is justified only if the fields (6.2) and (6.5) match along EH in the rates of deflection and rotation, that is, if

$$p(x_1,\ell/\sqrt{2}) = \bar{p}(x_1,\ell/\sqrt{2}) , \qquad \partial_2 p(x_1,\ell/\sqrt{2}) = \partial_2 \bar{p}(x_1,\ell/\sqrt{2}) , \qquad (6.6)$$

where $a = b = c = 0$ in (6.2). Since these conditions are fulfilled, we have found the collapse field for the optimal layouts.

In the corner triangle AEH, the principal rates of curvature are $q_1 = q_0$, $q_2 = -q_0$. Accordingly, one might expect beams in both principal directions. Since, however, q_2 changes from $-q_0$ to q_0 at the line EH, the bending moment Q_2 would have to vanish at this line. A beam in the x_2-direction would therefore have vanishing bending moment at AE or AH and at EH. Its bending moment would be positive in AEH, no matter how the load P is split between the beams in the two coordinate directions. Since this is not compatible with the negative q_2 , we only have beams in the x_1-direction.

In the central square EFGH, $q_1 = q_2 = q_0$, and any direction is a principal direction. Figure 6.1b thus shows an alternative optimal layout.

We now compare the weights of the grillages in Fig. 6.1 with the weight of a grillage whose beams are parallel to the sides of the square ABCD. At plastic collapse, the absolute value of the bending moment at any section of a beam in an optimal grillage equals the yield moment at this section. Moreover, the weight per unit length of a beam is proportional to its yield moment. The weight of an optimal grillage is therefore proportional to

$$\Omega = \int (|Q_1| + |Q_2|) \, dA , \tag{6.7}$$

where Q_1 and Q_2 are the bending moments in the beams of the grillage, dA is the area element of the planform of the grillage, and the integration is extended over this planform. The quantity Ω in (6.7) is called the mo-ment volume of the grillage.

For a collapse field with the rate of deflection p and the rates of cur-vature q_1 and q_2 , the equality of internal and external power of dissi-pation is expressed by

$$\int (Q_1 \, q_1 + Q_2 \, q_2) \, dA = \int P \, p \, dA , \tag{6.8}$$

where P is the load per unit area. For the collapse field of the optimal grillage, it follows from (6.1) that

$$Q_i \, q_i = \begin{cases} |Q_i| \, q_0 & \text{if } q_i = q_0 , \\ 0 & \text{if } |q_i| < q_0 , \end{cases} \right\} \quad \begin{array}{l} \text{no summation,} \\ i = 1, \, 2 . \end{array} \tag{6.9}$$

The left side of (6.8) thus equals $q_0 \, \Omega$, and we have

$$\Omega = q_0^{-1} \int P \, p \, dA . \tag{6.10}$$

Note that it follows from this relation that the grillage in Fig.6.1a has the same moment volume as that in Fig.6.1b, because both grillages cor-respond to the same collapse mechanism. For this mechanism, the rate of deflection is given by (6.2) with a = b = c = 0 in AEH and by (6.5) in EOH. For P = const, the moment volume (6.10) is found by integrating the-se rates of deflection over their areas of definition and multiplying the sum of the integrals by $4 \, P \, / \, q_0$. One finds

$$\Omega = 5 \, P \, \ell^4 \, / \, 6 . \tag{6.11}$$

While this evaluation of the moment volume is convenient, it is not constructive because it does not furnish any information concerning the yield moments $|Q_1|$ and $|Q_2|$, which determine the variable widths of the beams. We shall now evaluate the moment volume directly by determining the bending moments Q_1 and Q_2 . With respect to the axes shown in Fig. 6.1a, we have $Q_2 = 0$ in the corner triangle AEH, while Q_1 varies parabolically along the strip between $x_2 = x$ and $x_2 = x + dx$ with the maximum $Q_1 = P x^2 dx / 2$ at $x_1 = 0$ and $Q_1 = 0$ at $x_1 = \pm x$. The contribution of this strip to the moment volume Ω thus is $(2/3)(P x^2 dx/2)(2x) = 2 P x^3 dx/3$, while the contribution of the triangle AEH is

$$\Omega_1 = \int_0^{\ell\sqrt{2}} 2 P x^3 dx / 3 = P \ell^4 / 24 . \tag{6.12}$$

In the central square EFGH, we have beams in both coordinate directions. We shall attribute the load P/2 to each kind of beam. A strip parallel to EH with the width dx_2 then has the maximum bending moment $Q_1 = P \ell^2 dx_2/8$ at $x_1 = 0$ and $Q_1 = 0$ at $x_1 = \pm\ell/\sqrt{2}$. The contribution of this strip to Ω thus is $(2/3)(P \ell^2 dx_2/8)(2\ell/\sqrt{2}) = P \ell^3/(6\sqrt{2})$, while the contribution of all strips in the x_1-direction between EH and FG is

$$\Omega_2 = P \ell^4 / 6 . \tag{6.13}$$

Note that the beams in the x_2-direction between EF and HG make the same contribution to Ω .

Finally, we must consider the four edge beams of the square EFGH. Each of them carries a quarter of the total load on this square, that is, each of them carries the load $P \ell^2 / 2$ uniformly distributed over the span $\ell \sqrt{2}$, and therefore contributes

$$\Omega_3 = (2/3)(P \ell^2 / 2)(\ell \sqrt{2})^2 / 8 = P \ell^4 / 12 \tag{6.14}$$

to Ω . According to (6.12) through (6.14),

$$\Omega = 4\Omega_1 + 2\Omega_2 + 4\Omega_3 = 5 P \ell^4 / 6 \tag{6.15}$$

in accordance with (6.11).

The moment volume of a simply supported square grillage with beams parallel to the sides of the square may be obtained from (6.13), which gives
half this volume for a grillage with the side $\ell\sqrt{2}$. Accordingly, the desired moment volume for a square with the side 2ℓ is $4 \, P \, \ell^4 \, / \, 3$. The value in (6.15) is 37.5 % less than this.

6.2. Matching of basic types of collapse fields. By far the most fre-
quently encountered planform of grillages is the rectangle. This section
is therefore devoted to the matching of basic collapse fields near a
corner of a rectangular grillage. The edges forming this corner may be
simply supported or built in.

If a grillage in the first quadrant is built in along the edge $x_1 = 0$,
the rate of rotation $\partial_1 p$ and hence the rate of twist $-\partial_{12} p$ vanish along
this edge. The principal rates of curvature are therefore in the coordi-
nate directions, and $q_2 = 0$. This means that near the built-in edge $x_1 =$
0 , the beams are in the x_1-direction, and the rate of deflection is given by

$$p = q_0 \, x_1^2 \, / \, 2 \; . \tag{6.16}$$

If the edge $x_2 = 0$ is simply supported, the rate of deflection near this
edge must be of the form

$$\bar{p} = q_0 \, (\, a \, x_1 \, x_2 \, - \, b \, x_2^2 \, + \, c \, x_2 \,) \; . \tag{6.17}$$

Accordingly, the rates of curvature and twist in the coordinate directions are

$$q_1 = -\partial_{11}\bar{p} = 0 \; , \quad q_2 = -\partial_{22}\bar{p} = 2 \, b \, q_0 \; ,$$
$$\tag{6.18}$$
$$q_3 = -\partial_{12}\bar{p} = - \, a \, q_0 \; .$$

The principal rates of curvature thus are

$$\left. \begin{array}{c} q_{max} \\ q_{min} \end{array} \right\} = 0.5 \left\{ q_1 + q_2 \pm \sqrt{(q_1 - q_2)^2 + 4 q_3^2} \right\}$$

$$= \left\{ b \pm \sqrt{a^2 + b^2} \right\} q_0 . \tag{6.19}$$

Since q_{max} must have the value q_0 , we have

$$b = (1 - a^2) / 2 , \tag{6.20}$$

and hence

$$q_{min} = -a^2 q_0 . \tag{6.21}$$

It will turn out that $|a| < 1$. Accordingly, in the neighborhood of the simply supported edge $x_2 = 0$, all beams have the direction of q_{max}. If the angle this direction forms with the x_2-axis is denoted by θ , we have

$$\tan 2\theta = 2 q_3 / (q_1 - q_2) = a / b . \tag{6.22}$$

We must now see whether the fields (6.16) and (6.17) can be matched along a ray $x_2 = n x_1$ in the rates of deflection as well as the rates of rotation. In other words, can the parameters a, b, c, and n be chosen in such a manner that

$$p - \bar{p} = 0 , \text{ and}$$

$$(-n \partial_1 + \partial_2)(p - \bar{p}) = 0 \quad \text{along } n x_1 - x_2 = 0 ? \tag{6.23}$$

It is found that these conditions are fulfilled for

$$a = 1/\sqrt{2} , \quad b = 1/4 , \quad c = 0 , \quad n = \sqrt{2} . \tag{6.24}$$

It then follows from (6.22) that $\tan 2\theta = 2\sqrt{2}$ or

$$\tan \theta = 1/\sqrt{2} . \tag{6.25}$$

Other combinations of simply supported and built-in edges at corners of
rectangular plates can be discussed in a similar manner. Figure 6.2, in
which the letters S and B stand for these two kind of edge conditions,
shows the arrangements of beams near the corner. Full-line beam segments
have positive, and broken-line segments have negative curvature. With the
aid of the information in Fig. 6.2, optimal layouts for rectangular plates are readily obtained. For example, Fig. 6.3 shows an optimal layout

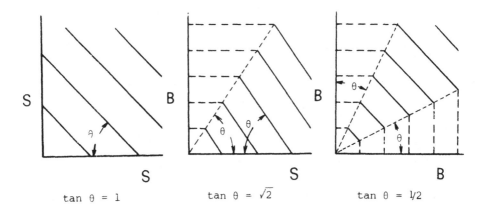

Fig. 6.2: Optimal layouts near corners

for a square plate that is built in on three edges and simply supported
on the fourth edge.

The optimal layout of beams near a corner formed by simply supported and
built-in edges is uniquely determined, but when one of the edges is free,
we do no longer have this kind of uniqueness. Figure 6.4, in which the
letter F indicates a free edge, shows optimal layouts for square grillages with three simply supported edges and one free edge (Fig. 6.4a) and
with two simply supported and two free edges (Fig. 6.4b). The way in

which the load is carried by the beams in Fig. 6.4a is obvious. In Fig. 6.4b, the load on the triangle ACD is completely carried by the beams parallel to AC. The load on the triangle ABC is carried by beams parallel to BD. The beam GHI, for instance, is simply supported at G and H, where it rests on the beam AC; it is free from load between G and H.

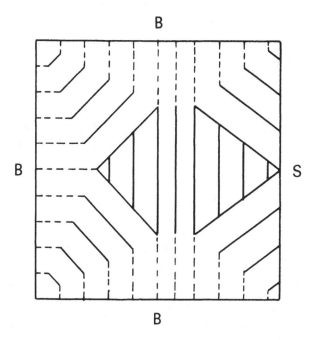

Fig. 6.3: Optimal square grillage with one simply
supported edge and three built-in edges

For further examples of rectangular grillages, the reader is referred to a paper by Rozvany[47]. Grillages with curved edges have been discussed by Morley[45], Sacchi and Save[48], and Rozvany[49]. Papers by Lowe and Melchers[50] though treating optimal design of fiber-reinforced plates of constant thickness are also relevant to the optimal design of grillages. An interesting method of grillage design, which was proposed by Heyman[51], does not necessarily furnish grillages of least weight.

In concluding this section, we briefly discuss the minimum-weight design
of a grillage of elastic beams for prescribed compliance to a given load.
The optimality condition for this problem stipulates a deflection field
with squares of the principal curvatures that do not exceed a given value
q_0^2 . Accordingly, the optimal layout of beams is the same as for the optimal plastic design considered above.

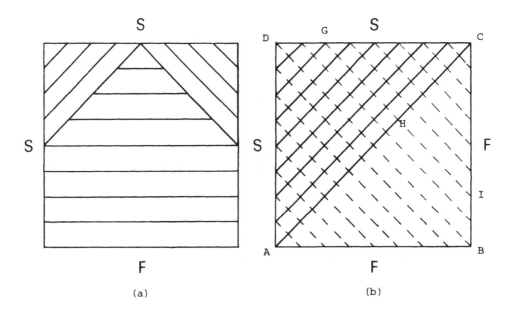

Fig. 6.4: Optimal grillages with one or two free edges

6.3. Two superposition principles. In this section, two superposition
principles will be discussed, which are useful in the optimal design of
grillages. The first of these concerns simultaneous loads, and the second
alternative loads.

It follows from (6.10) that, after division by q_0 , the rate of deflection field of the optimal grillage may be regarded as an influence field for the moment volume Ω . In other words, the optimal grillage for the simultaneous loadings P' and P" is obtained by the superposition of the optimal grillages for the single loadings P' and P". Here, the term superposition is meant to indicate that the yield moments of a beam element in the optimal designs for the single loadings P' and P" are added to obtain the yield moment of this element in the optimal design for the simultaneous loadings P' and P".

The superposition principle used in the optimal design for two alternative loadings P' and P" closely ressembles the principle in Section 5.2. In fact, proceeding as in that section, one readily shows that (5.13) is the optimality condition for the grillage under the alternative loadings P' and P", if q_i' and q_i'' stand for the rates of curvature of the beam element i in collapse mechanisms under the loadings P' and P", and q_0 is a constant reference rate of curvature. Following the proof in Section 5.2, one then finds that the optimal grillage for the alternative loadings P' and P" is obtained by the superposition of the optimal grillages for the single loadings $\bar{P} = (P'+P'')/2$ and $\bar{\bar{P}} = (P'-P'')/2$.

Figure 6.5 shows an example illustrating this kind of superposition (Rozvany[52]). A simply supported square grillage ABCD is to be optimally designed for the following alternative loadings: P' has the uniform intensity 2P and acts on the recangle AEFD, while P" has the same uniform intensity but acts on the rectangle EBCF. The loading $\bar{P} = (P'+P'')/2$ thus has the uniform intensity P and acts on the square ABCD, whereas $\bar{\bar{P}} =$ (P'-P")/2 has the uniform intensities P on AEFD and -P on EBCF. Figure 6.5a shows an optimal layout of the grillage for $\bar{\bar{P}}$, which is an alternative to the layouts in Fig. 6.1. Since the loading $\bar{\bar{P}}$ is antisymmetric with respect to the line EF, the rate of deflection will vanish along this line. In the rectangle AEFD, the optimal layout for $\bar{\bar{P}}$ thus corresponds to simple support along all edges of this rectangle, and a similar

remark applies to the rectangle EBCF. Figure 6.5b shows an optimal lay-
out for $\overline{\overline{P}}$. The yield moments of the beams of the component grillages

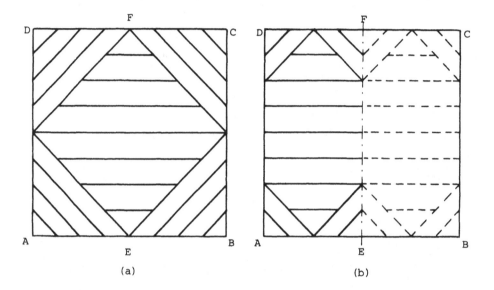

Fig. 6.5: Component grillages of optimal grillage for alternative
loadings of uniform intensity P acting on AEFD or EBCF

in Fig. 6.5 are readily determined, and the optimal grillage for the alternative loadings P' and P" is obtained by the superposition of these
component grillages.

REFERENCES

1. J. Farkas, J. reine & angew. Math. $\underline{124}$, 1 (1902)
2. C. Y. Sheu & W. Prager, J. Optim. Theory & Appls. $\underline{2}$, 179 (1968)
3. J.-M. Chern & W. Prager, J. Optim. Theory & Appls. $\underline{5}$, 424 (1970)
4. J. B. Martin, J. Optim. Theory & Appls. $\underline{6}$, 22 (1970)
5. J.-M. Chern & J. B. Martin, Z. angew. Math. & Phys. $\underline{22}$, 834 (1971)
6. W. Prager & J. E. Taylor, J. Appl. Mech. $\underline{35}$, 102 (1968)
7. F. I. Niordson, Quart. Appl. Math. $\underline{23}$, 47 (1965)
8. M. Turner, AIAA J. $\underline{5}$, 406 (1967)
9. J. E. Taylor, AIAA J. $\underline{5}$, 1911 (1967) and $\underline{6}$, 1379 (1968)
10. C. Y. Sheu, Int. J. Solids & Structs. $\underline{4}$, 953 (1968)
11. B. R. McCart, E. J. Haug, & T. D. Streeter, AIAA J. $\underline{8}$, 1012 (1970)
12. M. S. Zarghamee, AIAA J. $\underline{6}$, 749 (1968)
13. B. L. Karihaloo & F. I. Niordson, J. Optim. Theory & Appls. $\underline{11}$, 638 (1971)
14. D. C. Drucker & R. T. Shield, Proc. 9th Int. Congr. Appl. Mech., Brussels, $\underline{5}$, 212 (1957)
15. J. E. Taylor, J. Appl. Mech. $\underline{34}$, 486 (1967)
16. J. E. Taylor & C. Y. Liu, AIAA J. $\underline{6}$, 1497 (1968)
17. W. Prager, Z. angew. Math. & Phys. $\underline{19}$, 252 (1968)
18. L. J. Icerman, Int. J. Solids & Structs. $\underline{5}$, 473 (1969)
19. Z. Mróz, Z. angew. Math. & Mech. $\underline{50}$, 303 (1970)
20. R. H. Plaut, Quart. Appl. Math. $\underline{29}$, 315 (1971)
21. R. T. Shield & W. Prager, Z. angew. Math. & Phys. $\underline{21}$, 513 (1970)
22. J.-M. Chern & W. Prager, J. Optim. Theory & Appls. $\underline{6}$, 161 (1970)
23. J.-M. Chern, Int. J. Solids & Structs. $\underline{7}$, 373 (1971)
24. W. Prager, Int. J. Mech. Scis. $\underline{12}$, 705 (1970)
25. Z. Mróz, Archiwum Mech. Stosow. $\underline{15}$, 63 (1963)
26. W. Prager, Proc. Nat. Acad. Scis. $\underline{61}$, 794 (1968)
27. W. Prager, contrib. to "An Introduction to Structural Optimization", University of Waterloo, Ontario, Canada, 165 (1969)
28. Z. Mróz, Z. angew. Math. & Mech. $\underline{50}$, 303 (1970)
29. W. Kozlowski & Z. Mróz, Int. J. Solids & Structs. $\underline{12}$, 1007 (1970)
30. P. V. Marçal & W. Prager, J. de Mécanique $\underline{3}$, 509 (1964)
31. W. Prager & R.T. Shield, J. Appl. Mech. $\underline{34}$, 184 (1967)
32. D. E. Charrett & G. I. N. Rozvany, Int. J. Non-linear Mech. $\underline{7}$, 51 (1972)
33. M. A. Save, J. Struct. Mech. $\underline{1}$, 267 (1972)
34. M. R. Horne, J. Instn. Civil Engrs. $\underline{34}$, 174 (1950)
35. G. I. N. Rozvany, J. Appl. Mech. $\underline{41}$, 309 (1974)
36. W. Prager & P. G. Hodge, Jr., Theory of Perfectly Plastic Solids, New York, 1951
37. A. G. M. Michell, Phil. Mag. (6) $\underline{8}$, 589 (1904)
38. R. Hill, The Mathematical Theory of Plasticity, Oxford, 1950
39. W. Prager, Transactions, Royal Institute of Technology, Stockholm, Nr. 65 (1953)

40. G. A. Hegemier & W. Prager, Int. J. Mech. Scis. 11, 209 (1969)

41. W. S. Hemp, Optimum Structures, Oxford, 1973, p.25

42. J. C. Nagtegaal & W. Prager, Int. J. Mech. Scis. 15, 583 (1973)

43. W. Prager, Comp. Methods Appl. Mech. & Engg. 3, 349 (1974)

44. A. C. Palmer & D. J. Sheppard, Proc. Instn. Civil Engrs. 47, 363 (1966)

45. C. T. Morley, Int. J. Mech. Scis. 8, 305 (1966)

46. G. I. N. Rozvany, J. Amer. Concrete Inst. 63, 1077 (1966)

47. G. I. N. Rozvany, Comp. Methods Appl. Mech. & Engg. 1, 253 (1972)

48. S. Sacchi & M. Save, Int. Assoc. Bridge & Structl. Engg. 29-II, 157 (1969)

49. G. I. N. Rozvany, Int. J. Mech. Scis. 14, 651 (1972)

50. P. G. Lowe & R. E. Melchers, Int. J. Mech. Scis. 14, 311 (1972), 15, 157 and 711

51. J. Heyman, Proc. Instn. Civil Engrs. 13, 339 (1959)

52. G. I. N. Rozvany, Proc. IUTAM Symposium on Optimal Structural Design (Warsaw, 1973), to appear

Printed in the United States
By Bookmasters